The Field Naturalist

W.A. WAISER

The Field Naturalist

*John Macoun,
the Geological Survey,
and Natural Science*

UNIVERSITY OF TORONTO PRESS

Toronto Buffalo London

© University of Toronto Press 1989
Toronto Buffalo London
Printed in Canada

ISBN 0-8020-2686-9

Printed on acid-free paper

Canadian Cataloguing in Publication Data

Waiser, W.A.
The field naturalist
Includes bibliographical references and index.
ISBN 0-8020-2686-9
1. Macoun, John, 1831–1920. 2. Geological Survey
of Canada – History. 3. Victoria Memorial Museum
(Canada). 4. Naturalists – Canada – Biography.
I. Title.
QH31.M15W35 1989   574'.092'4   c88-094902-3

This book has been published
with the help of a grant from the
Canadian Federation for the Humanities,
using funds provided by the Social Sciences and
Humanities Research Council of Canada.
Publication has also been assisted by
the Canada Council and Ontario Arts Council
under their block grant programs.

for Jean Lois Ritchie

Contents

Preface

In November 1881, Professor John Macoun of Belleville, Ontario, was appointed to the Geological Survey of Canada as a political reward for his favourable assessment of the agricultural potential of western Canada during the 1870s. Over the next thirty years, first in his capacity as Dominion botanist and then as Survey naturalist, Macoun scorned the prevailing trend towards specialization in a particular area of biology in favour of collecting as widely and thoroughly as possible. This concentration on field work had its costs and benefits. Although Macoun always returned from the field with large collections, the Survey's natural history work suffered from the attempt to do too much. Essential field data were not always recorded with the specimens, while the collections themselves were sometimes poorly prepared or not catalogued until several years later. It also became necessary for Macoun to rely heavily on foreign, mostly American, natural scientists to identify his difficult specimens and ensure that they were accurately named. Macoun's collecting activity, at the same time, was nothing short of phenomenal in terms of area covered and species discovered. His field work, despite its shortcomings, also helped lay the foundation for the National Museum of Canada. Ironically, though, when the new Victoria Memorial Museum was finally occupied in 1912, the work of the biological division remained rooted in the nineteenth century.

A number of individuals and institutions should be thanked for their assistance in the preparation of this study. Scientists at the National Museums of Canada were most co-operative in allowing me to poke around various buildings and rummage through storage closets and old filing cabinets. The staff of the Government Archives Division, National Archives of Canada, in particular Terry Cook and former archivist Robert Hayward, were extremely generous with their time, guidance, and interest. Robert Hayward and Kate Fawkes kindly provided a home away from home during my visits to Ottawa. Macoun descendants Eleanor Sanderson and James Macoun Kennedy helped in whatever way they could. Jack Cranmer-Byng, who is currently preparing a biography of Percy Taverner, directed me to the Royal Ontario Museum material and sent me a number of important letters.

This study originated as a doctoral dissertation under the able supervision of T.D. Regehr. My editor, Gerry Hallowell, expressed an interest in the manuscript at an early date and was a constant source of guidance and encouragement. John Warkentin of York University read and provided expert comment on chapter one. My good colleague J.R. Miller went over the manuscript with great care and attention and made a number of excellent suggestions – as did the anonymous readers for the University of Toronto Press and the Canadian Federation for the Humanities. Naturally, any errors of fact or interpretation are my own. The typing of the manuscript at various stages was handled by Jean Horosko in her usual efficient and tireless manner. Maragaret Allen did a fine job as copy-editor.

The J.S. Ewart Fund of the University of Manitoba financed a research trip to Ottawa, while the College of Arts and Science, University of Saskatchewan, greatly facilitated the writing of the manuscript through the award of a release time grant. The two maps were prepared by cartographer Keith Bigelow of Saskatchewan's Department of Geography. Funding for the maps and photographs was provided by the President's Publication Fund, University of Saskatchewan.

Finally, Marley, Jess, Mike, and Katie deserve a special word of thanks for their help along the way. It's time for a few field trips of our own.

John and Ellen (Terrill) Macoun on their wedding day,
1 January 1862

Professor Macoun with young naturalist
in the Maritimes in the late 1890s.
The Professor's unrivalled prowess as a field naturalist was recognized in 1948
with the establishment of the Macoun Field Club in Ottawa.
Co-sponsored by the Ottawa Field-Naturalists' Club
and the National Museum of Natural Sciences,
it encourages a love of wildlife and concern for conservation
among young naturalists.

Seventy-year-old John Macoun in January 1902.
He would spend that summer investigating the agricultural potential
of the Klondike.

A young James Macoun (foreground)
during one of his first Survey expeditions
as a member of the 1885 A.P. Low expedition to the
Lake Mistassini region of central Quebec.

James Macoun en route to Hudson Bay in 1910.
He was shipwrecked at Wager Inlet
but still managed to return with nearly 200 pounds of
natural history specimens.

William Spreadborough (1856–1931),
probably on his wedding day.
The Macouns' field assistant for thirty years,
he repeatedly declined to take part in any
collecting work after their deaths.

Dr A.R.C. Selwyn was impressed by Macoun's ability
to apply his skills as a plant geographer to the
nation-building tasks at hand
and invited him to join his 1875 GSC expedition
to the Peace River country.

Kicking Horse Pass, 12 September 1884.
During the 1884 British Association for the Advancement of Science tour
of western Canada, a geological party under Dr A.R.C. Selwyn
escaped near disaster when the Survey director's
hammering at the entrance of a railway tunnel in the Kicking Horse Pass
caused a rock fall.

Dr R.W. Brock, Survey director (1908–14),
wanted to end the wide-ranging, all-inclusive field surveys of the past
in favour of greater specialization in the various areas of Survey endeavour.
Thanks to Macoun's influence, however, the work of the
natural history branch remained rooted in the nineteenth century.

Dr George Mercer Dawson,
Geological Survey director (1895–1901),
died before knowing that his efforts to secure a new home for the Survey
and its museum had been successful.

Acting Survey director (1901–6) Robert Bell
was held responsible for the publication of
James Macoun's controversial 1904 Peace River report.

Professor Macoun at work in his new office
in the Victoria Memorial Museum, 1911.
Rows of herbarium cases line the wall to his back – an improvement
over the Survey's old building where they were scattered
up and down the corridors.

Professor Macoun,
flanked by 'Bugs' Young (*left*) and William Spreadborough,
at their collecting base at Ucluelet,
Vancouver Island, 1909

John Macoun and Percy Taverner
on a collecting trip in 1911.
Taverner was amazed by the octogenarian's energy:
'Works through swamp, thicket and meadow with all the
energy of a young man. He is a wonder.'

Percy Algernon Taverner.
Taverner joined the Survey's natural history staff in May 1911
as assistant naturalist and curator in charge of vertebrates.
Although the Macouns had just published a revised bird catalogue in 1909,
Taverner immediately started work on a new bird book of his own.

Rudolph Martin Anderson, Mackenzie Delta area, 1914.
Anderson, an American-born and trained mammalogist,
served as southern leader of the Canadian Arctic Expedition (1913–18).
He assumed the position of chief of the Geological Survey's
biological division upon James Macoun's death.

Hiking party, August 1912.
Left to right: James Macoun; A.O. Wheeler, James' brother-in-law;
L.C. Wilson; Oliver Wheeler, A.O. Wheeler's son; A.R. Hart.

Dominion horticulturalist W.T. Macoun visits his parents at Sidney.
Standing left to right: W.T. Macoun, Nellie (Macoun) Sanderson, John Macoun;
seated: Ellen Macoun, grandson John Macoun, and his mother,
Elizabeth Macoun.

OPPOSITE

Ellen Macoun and Miss W.H. Fatt, Sidney, Vancouver Island, 1920.
Miss Fatt served as Professor Macoun's secretary during the preparation
of his autobiography.

A social gathering at the A.O. Wheeler home at Sidney, Vancouver Island.
John and Ellen Macoun are on the far left,
while daughter Clara and her husband, A.O. Wheeler,
are seated in the middle.

The Geological Survey of Canada's
Sussex Street headquarters in lower town, Ottawa.
The former Clarendon House, the building proved inadequate
from the time it was occupied by the Survey in 1881.

OPPOSITE

Exhibition Hall, Survey museum,
Sussex Street headquarters.
Large geological specimens had to be kept in the basement
because of the structural weakness of the building.

The Victoria Memorial Museum (September 1907)
proved a construction nightmare.
The site that the Laurier government had taken so long to decide upon
was found to be underlain with unstable clay.

Victoria Memorial Museum, 1912.
The department of public works originally wanted to create
a grand promenade along Metcalfe Street between Parliament Hill
and the new museum, similar to that which existed in Washington along
Pennsylvania Avenue between the White House and the Lincoln Centre.

Victoria Memorial Museum, 1927.
The central or 'Laurier Tower,' as it was sarcastically dubbed,
had been removed in 1915 – it fell down once during construction
and the department of public works was afraid that it might do so again.
In 1927 the museum was officially declared the
National Museum of Canada.

The Field Naturalist

Introduction

When Canada took possession of Rupert's Land and the Northwestern Territory on 15 July 1870, the scientific community of the young Dominion was presented with a wonderful opportunity. Just eleven years earlier, the celebrated British naturalist Charles Darwin had shaken the scientific world with the publication of his *Origin of Species*, in which he boldly argued that variation in nature was the result of the constant adaptation of species to their environment. This theory of 'natural selection' initiated an intensive world-wide search for new species, as naturalists attempted to find clues to the relationship between variation and speciation. In Canada, the newly acquired western interior, with its diverse wealth of plant and animal life, could have served as an important testing ground for Darwin's hypothesis. It could have been a vast natural-history laboratory. Yet apart from the vigorous criticism of a few Canadian intellectual leaders, Darwin and his ideas were not welcome in Canada.[1] Naturalists regarded the natural world as the handiwork of the all-wise Creator; species were fixed and unalterable. The possibility of testing evolution on the plains, parklands and forests was thus never pursued – probably never even seriously considered – by Canadian natural scientists. Natural history did nonetheless play a significant role in the settlement of the region, but on practical not abstract grounds.

Canadian acquisition of the North-West was predicated on the twin assumptions that the land had great promise as an agricultural frontier and that development of the region would be the means to empire.[2] Successive federal administrations consequently devoted a considerable amount of energy to the opening of the region. In particular, Ottawa expected government-sponsored scientific surveys to provide the kind of practical, positive information that would facilitate exploitation of the west's resources. It did not believe in science for its own sake but rather in the use of science to promote the material interests of the country.[3] One naturalist who played a dominant role in the provision of this practical knowledge in the late nineteenth and early twentieth centuries was the botanist John Macoun.

John Macoun,[4] or 'the Professor' as he was popularly known, personified Canadian natural history in the mid-nineteenth century. Born in the parish of Maralin, County Down, in northern Ireland, on 17 April 1831, John was raised on family land that had been granted to one of his father's ancestors almost two centuries earlier for military service during the English Civil War. It was an ideal setting for a young boy blessed with a child's insatiable curiosity, and he developed a great passion for the outdoors and the natural world around him. He spent as much time chasing sticklebacks in the local streams or counting birds' nests in the orchards as he did helping in the fields and meadows. And whenever possible, he would steal away to one of his favourite hiding places for several hours, often returning with an armful of wildflowers for his garden.

Macoun's Ulster childhood also had a profound influence on his character and values. In October 1837, his father, a retired soldier who had helped put down the Irish Rebellion of 1798, died, leaving his mother with four young dependants – a girl and three boys. Although relatives rallied to the aid of the family, John, the second youngest, learned from an early age 'that the world gave more blows than favours.'[5] He became exceedingly stubborn, almost self-righteous, in his determination to succeed and never shied away from a challenge, no matter how unfavourable the odds. 'I will always maintain that the man who fights the longest wins,' he told an Ottawa audience later in his life. 'It is the chap who stops first who loses the battle.'[6] His education at the parochial school of the Presbyterian church not only strengthened this profound belief in his capabilities but gave it moral overtones. He came to pride himself on telling the truth, even if it meant a whipping, and took a temperance pledge that he honoured for the rest of his life. His schooling also instilled in him a sense of superiority and, although his mother tried to shield him from the sectarian strife of the region, he took great delight in fighting the local Catholic youths. By the time he assumed a clerk's position in Belfast in his early teens, Macoun held the same values as his fellow Ulstermen – allegiance to the British Crown, dedication to the union with Great Britain and the wider empire, and support for the Conservative party and the anti-Catholic Orange Order.

In the spring of 1850, the Macoun family decided to sell their land and, like tens of thousands of other Irish immigrants during this period, seek the promise of a materially better life in Canada. Settling near their mother's brother in Seymour Township, Northumberland County, about ninety miles east of Toronto, they were disappointed to find that the region was still in a primitive stage of development and that cleared land was both scarce and expensive. The family had enough money to buy a hundred-acre farm and, when not clearing their own land, the three brothers were often earning extra income by working for local farmers. John was initially inept at these tasks, but he learned quickly and soon applied for his own homestead. And whether felling trees or clearing stones, as he had during his youth he always took time to investigate the local flora. This interest in the native plant life developed into a serious study of Canadian botany with the encouragement of a local farmer who provided Macoun with an old published list of English plants based on the Linnaean classification system. Using this guide and any other books that he could get his hands on, he would pick a plant at random and attempt to identify it. It was a slow, painstaking process, but Macoun's enthusiasm never waned.

In 1856, Macoun decided to change occupations. Although he had held his land for five years, he was never satisfied as a simple, backwoods farmer; the only source of contentment and purpose in his life at this time was his continuing fascination with the natural world. Macoun therefore decided to become a public school teacher on the assumption that it would enable him to devote more time to study and collecting. Education in Canada West during the 1840s and 1850s was undergoing a fundamental transformation as the number of schools and pupils increased and the quality of teaching was improved through the establishment of a uniform system of examination and certification. In frontier areas, however, good teachers were difficult to retain, particularly because of the low salaries they were paid.[7] When Macoun consequently went to be examined on the strength of a few days' grammar study, he encountered a sympathetic inspector who, after a brief, informal discussion in his cutter, recognized that the young man had potential and granted him a teaching certificate, provided he serve a kind of three-week apprenticeship in the local school.

Macoun spent the first two and a half years of his new career in Brighton, a small village west of Belleville. Short, slight in build, and now sporting a beard that was characteristic of many of his generation, he had been hired by a somewhat reluctant school board; it was not often that someone came directly off the farm and into the classroom. The school officials were struck, however, by his enthusiasm – a quality that impressed most people on first meeting – and they decided to give him a chance. They had nothing to worry about. An engaging teacher, Macoun tried to make his classes as stimulating as possible, while always encouraging his students to ask questions. 'Enthusiasm wins every-time,' he once advised a young colleague, 'and only those who are full of it rise to the top and stay.'[8] When not reading or preparing lessons, he continued to collect and take notes on plants. This activity, particularly the long hours he spent scrutinizing specimens at his boarding house, mystified his future father-in-law, who believed the young man should be paying more attention to his fiancée.

In the interests of broadening his subject knowledge and improving his teaching skills, Macoun attended the 1859 fall session of the Toronto Normal School. These were some of the happiest days he had known since his arrival in Canada. He had never seen such a large school, let alone heard a lecture before, and he thrived on the collegial atmosphere. He found that his studies greatly benefited from the reading he had done on his own and concluded that 'independent thought was the power that always won.'[9] He also became good friends with a university student who was amused by his backwoods botany and tried to teach him something about classification and structure during their regular Saturday outings together.

Upon graduating at Christmas with a first-class certificate, Macoun began teaching, as previously arranged, in the small village school at Castleton, Northumberland County. It was a modest posting for the recent graduate, but at least he was near his family. More important, the fields and forests were close at hand, and he often slipped away for a few hours collecting with the local doctor, another botanist. Working without the aid of a microscope or magnifying glass, Macoun had to identify many of his specimens with the naked eye – a skill he perfected during his rambles in the field. He believed that 'a good collector of anything must have a trained eye and besides that must only *see* things he was searching for.'[10] There were a growing number of plants, however, that he could not make out.

In the fall of 1860, Macoun applied for a vacant position in number one school, Belleville, and was hired by the owner and editor of the *Belleville Intelligencer* (and future Conservative prime minister), Mackenzie Bowell. Belleville, a thriving community of more than 5,000, embodied the same kind of restless ambition and bold self-assurance that fuelled Macoun. Located on the Bay of Quinte at the mouth of the Moira River, the Loyalist-founded town was a major centre for lumbering and mixed farming. And with its recent designation as a divisional point on the Grand Trunk Pacific rail line between Toronto and Montreal, Belleville was confident that it would soon emerge as a major commercial centre.[11] Macoun, for his part, rapidly became one of the city's most popular teachers. Drawing upon his Normal School experience, he taught science as a regular curriculum subject, believing that an emphasis on classics alone was outdated. He also employed the 'object lesson' method of teaching and was forever devising practical illustrations for his students' benefit instead of forcing them to memorize material from textbooks. '*Make* and *mount* a good collection,' he told a fellow teacher in reference to his natural history lessons, 'and a textbook will *not be a necessity* but your knowledge will be transmitted to others by actual (teaching) converse.'[12]

Macoun's life also underwent a number of personal changes in Belleville. On 1 January 1862, he married Ellen Terrill, a quiet, mild-mannered woman of Quaker background whom he had been courting since he began teaching in Brighton. Later that fall, the small house that he had bought behind Belleville's Albert College echoed to the sounds of the first of five children, James Melville. Probably through his growing friendship with Mackenzie Bowell, Macoun also became an active member of St Andrew's Presbyterian Church and a supporter of the Orange Lodge and the Conservative party. In fact, drawing upon the values of his Ulster childhood, he took an ardent interest in the movement towards Confederation and the promised benefits of union. 'I see a glorious future dawning on my adopted country,' he observed in October 1864. '[W]e love the British connection, we love our Queen and despise the Yankee.'[13] Two years later, he put his words into action when he served briefly as a volunteer at Prescott during the Fenian troubles.

His professional and personal life aside, Macoun looked upon his move to Belleville as a major turning point, for he decided to devote every spare moment to the study of botany. Rising at four a.m. in summer and making large collections before breakfast, he rapidly developed an expert knowledge of the range and habitat of many plant forms. He also started a private herbarium in his new home in 1862 and set his sights on assembling a representative collection of Canadian flora. Macoun was convinced that the development of a naturalist had to be based primarily on work in the field. 'Text-book natural history is of small account,' he once advised an aspiring collector in a characteristic swipe at the Canadian academic community, 'unless supplemented by practical work in the field and how very, very few of our college professors are field men.'[14]

This intensive collecting activity came on the heels of the publication of one of the most important books of the nineteenth century, Charles Darwin's *Origin of Species*. Up until this time, although there had been periodic questioning of the fixity of species, the natural world was traditionally regarded as the handiwork of God, in which every member of the plant and animal kingdom was a fixed and stable species, occupying a special place in the divine scheme of things. There was no conflict between science and religion; they were regarded as one and the same. The study of nature was the study of God's creations.[15]

Darwin's book represented a fundamental challenge to this prevailing conception of the natural world, because it focused attention upon interrelationships within populations and changes within species. According to the British naturalist's theory, every species produces more offspring than a particular environment can support. In the ensuing struggle for food and living space, those individuals that happen to have some slight advantage over others will survive. It might be a matter of a slightly curved beak or lighter colouring, but whatever the difference, those individuals will live to perpetuate their distinctive character traits. This idea of 'natural selection' was disturbing, in that it suggested that the supposedly constant and unalterable creations of God were continually adapting to their changing environment. The world of nature was no longer harmonious in the sense of unchanging. Equally unsettling was the notion that this adaptive selection in nature was the work of nature itself. God had been bumped from the stage of life and relegated to a seat in the rear balcony.[16]

The *Origin of Species* provoked a storm in religious and scientific circles in Great Britain and the United States. In Darwin's homeland, fellow naturalist Thomas Huxley, although believing that the concept of natural selection failed to explain the actual cause of variation within species, provided a spirited defence in the face of fierce opposition. Asa Gray, rapidly emerging as America's pre-eminent botanist, was also concerned about the mechanism of variation but represented the Darwinian side in a series of famous debates in the United States with Louis Agassiz, curator of the Museum of Comparative Zoology at Cambridge. Darwin's controversial work also had a profound effect on natural history and eventually led to the emergence of the new science of biology. At universities in Europe and the United States, natural scientists, especially those trained in medicine, began to focus on the dynamic relationship between structure and function. The whole life history of organisms took on new meaning, causing many investigators of the natural world to forsake the field and the collecting basket in favour of the laboratory, the microscope, and the experiment.[17] This trend did not mean, however, that the collection of specimens became any less important. Taxonomic data could help resolve the question of natural selection by providing clues to variation among species. The Darwinian revolution therefore initiated a great period of exploration for new species on a world-wide scale, while bringing about classification adjustments that better reflected the order of nature.[18]

Natural history enjoyed a widespread popularity in British North America in the 1860s, as evidenced by the growing number of natural history societies and the swelling ranks of amateur and professional collectors from coast to coast. The motivating factor behind this flurry of activity, however, was not Darwin's speculations. A few Canadians expressed interest in the new theory, among them geologist and palaeontologist William Dawson, principal of McGill University; but Darwin's ideas were alien to the Canadian tradition of natural science, and thus were largely ignored.[19] Nature study in Canada in the mid-nineteenth century was essentially Baconian in method, concentrating on the careful, methodical observation, listing, and description of God's workmanship and eschewing abstract theory building. As such, not only was it something that everybody – layman and professional alike – could take part in, but also it offered the possibility of making discoveries new to science, particularly in light of Canada's relatively unexplored natural history frontiers. This sense of common purpose also facilitated communication between amateur and professional scientists; the exchanging of lists and to a lesser degree specimens characterized scientific correspondence during this period.[20]

Science in Victorian Canada was also regarded as a 'useful' activity, entirely in keeping with the Victorian character. It was an uplifting exercise that required stamina, self-discipline, industry, and efficiency. Field work also led to an intimate knowledge of nature's many forms – a knowledge that was necessary to discover the possible uses to which Canada's resources could be put. There was a great faith in the problem-solving capabilities of science in Victorian Canada; scientific knowledge was equated with with industry, progress, and power.[21]

Victorian Canadians firmly believed, then, that religion and nature were inextricably related and that the study of nature's various forms constituted a form of worship of God's wondrous bounty – a bounty that had been specifically provided for man's use and benefit. This God-centred philosophy of nature lay at the heart of the Canadian educational system at mid-century and was adhered to by many of the country's leading scientists and scholars.[22] In an 1857 article on the role of nature study, for example, McGill botanist James Barnston (eldest son of noted botanist George Barnston) described how 'the general contemplation of the works of Creation ... exercise[s] a wholesome influence on the mind, excite[s] admiration within the breast and encircle[s] the imagination with a halo of pleasurable feelings.'[23] But he also warned that mere observation 'is not all that is required of man, to whom was given *power*, to have command over "things possessing life," and *intelligence*, to study with advantage to himself the numerous and varied objects placed before him.'

Macoun's intimate knowledge of the living plant and its habitat put him in an ideal position to appreciate Darwin's work. The British scientist was challenging botanists to look beyond the dried herbarium specimen towards the environmental conditions under which plants flourished and adjusted. Yet as a devout Presbyterian, Macoun could not understand how any observer of nature could doubt God's creative role. From the very beginning of his interest in the natural world, he regarded each field trip as a source of emotional satisfaction and spiritual solace – 'in the realm of Nature, God's hand was ever open to strew one's path with beauty and fill one's heart with praise.'[24] This thinking was confirmed by his reading during these years. In *Principles of Geology*, the eminent British geologist Charles Lyell argued that species were fixed and stable entities that had been purposefully created to exist under a particular set of ecological conditions. A change in these conditions, according to Lyell, would cause species to become extinct; a species could not be altered. Scottish geologist Hugh Miller echoed these ideas, suggesting that the geological record was not in conflict with Scripture, in particular the six days of Creation, and that complicated life forms were found in the early fossil record.[25] In light of these readings, his own religious views, and the general dismissal of Darwin in Canadian scientific and academic circles, Macoun never gave much thought to the 'Development Theory.' 'Believe me,' he told one of his early mentors, 'my principles are too well grounded for me to become a *disciple* of a school that seems to ignore the fact that God is the author of all being.'[26] He briefly expanded upon this attitude in the only other known reference to Darwin's work in his voluminous correspondence: 'Mr. Darwin is most likely right in his opinion but I doubt it. Habit is everything in nature or rather instinct. Nature's loves are very simple. We are all becoming so refined that in many cases we discard them and conjure up theories of our own.'[27]

Despite this flat rejection of Darwin's ideas, Macoun was not a Baconian drudge – a mere descriptive naturalist who eschewed all generalizations. In addition to reading Lyell and Miller, he also purchased Alexander von Humboldt's monumental classic on physical geography, *Cosmos*. Humboldt called for the detailed, firsthand measurement of diverse yet interrelated natural phenomena in the search for general laws or dynamic explanations. In particular, he attributed the range and distribution of plant and animal life to the interaction of climate and terrain.[28] Humboldt's work had a critical influence on Macoun's field methods, for the amateur botanist was soon 'ready with an answer for almost any natural cause, right or wrong.'[29] It was also on the basis of his reading of Humboldt that Macoun probably developed his method of assessing the agricultural capabilities of a region. He was convinced that the natural flora of a district indicated the character of the soil and climate and hence the suitability of the region for cultivation purposes. He later boldly claimed that '[T]his botanical test was the only true criterion by which the agricultural status of any district should be judged.'[30] Ironically, Darwin had been severely criticized by his Canadian opponents for making similarly daring generalizations. Macoun's case was fundamentally different though, in that he regarded his field efforts as an attempt to uncover God's ordered universe. His ability, moreover, to go beyond the simple descriptive listing of nature's bounty and apply his botanical skills to practical ends not only suited Canadian thinking on nature study but made his scientific work particularly attractive.

Macoun's field work and study in the early 1860s established his standing as an expert on the botany of southeastern Ontario, and he began receiving visits from such noted Canadian botanists as George Barnston, Abbé Louis-Ovide Brunet, and George Lawson. In December 1860, Lawson had been instrumental in the formation of the Botanical Society of Canada, an organization designed to facilitate the growth of botanical knowledge and its practical applications,[31] and he invited the collector to attend the next meeting in Kingston. Although as a mere schoolmaster Macoun initially felt out of place, his outspoken manner and enthusiasm for the subject led to a number of friendships with established botanists and aspiring students. Macoun also earned a reputation in Great Britain and the United States as a first-rate collector – a reputation that he personally generated. Lacking any formal botanical training, he found that he could not name several of his specimens and consequently availed himself of the knowledge and assistance of a number of prominent British and American specialists. In striking up a correspondence with the noted British botanist Sir William Hooker in November 1863, for example, he boasted that the herbarium that he had started almost a year earlier already contained 1,100 different species and was the largest private collection in the province.[32] Sir William, who was considering the publication of a colonial flora for Canada, overlooked such cockiness and helped the amateur collector resolve his identification problems in exchange for a set of plant duplicates. This assistance, along with that of a number of respected American botanists who were equally anxious to examine his specimens, was interpreted by Macoun as an endorsement of his collecting prowess.[33] It also had a tremendous impact on his future scientific career, in that a pattern was established whereby Macoun tended to concentrate on his field activities and leave the critical determination of any questionable specimens to his correspondents. It was pioneering work, nonetheless, for in 1864, the Reverend Chester Dewey, a prominent local botanist in Rochester, New York, named a new species of sedge, *Carex macounii*, after the Belleville botanist. This honour would be repeated many times throughout Macoun's life.

Anxious to collect in different environments, Macoun embarked on his first major field trip in July 1865 up the Hastings Road, a settlement trail that cut through the length of the township to the rugged forested uplands along the edge of the Canadian Shield. It was a highly rewarding trip that resulted in 'many rare and interesting things,' including '32 species of flowering plants which I had not observed before in Canada.'[34] Such success, however, only whetted Macoun's botanical appetite for more and led to frustration. What with his teaching duties, limited salary, and growing family – a daughter, Clara, was born in 1864 – he was effectively prevented from undertaking the wide-ranging·botanical surveys that he was certain would yield species new to science and thereby enhance his standing in the scientific community. 'I sometimes lament my poverty,' he confided to Dewey, 'that prevents me from following up my success in making such discoveries in Canadian botany as I have done over the last five years ... I wish I had a wider field.'[35] Dewey, who was something of a father figure to Macoun, suggested that the Geological Survey of Canada might be interested in the collector's talents and contacted the agency on his behalf. Although nothing came of Dewey's inquiry, the idea of being attached to the government agency was attractive to Macoun and he never gave up on the notion. The following year, when he learned that A.T. Drummond, a fellow botanist and one of Lawson's former students at Queen's, had accompanied a Survey expedition, he was quite jealous. 'I wish I had got the chance from Sir William Logan [Survey director] you did. I would have jumped at it.'[36] He also issued a challenge. 'Although I say it,' he bragged to Drummond, 'I can collect more species on a given area than any other person in Canada. And why? because I know exactly where to look for them.'[37]

Macoun's situation brightened considerably in 1868 when he was approached by Albert Carman, principal of Albert College, and asked to assume a new chair in natural history. Just two years earlier, the college had received its university charter and in keeping with its new status, Carman was anxious to expand the teaching staff in previously neglected areas. Not just any naturalist would do, though. A traditional Methodist, Carman was probably attracted to Macoun because they shared the same views on nature; he hired Macoun to teach botany, geology, and theology.[38] To ensure that there was no questioning of Macoun's academic credentials, moreover, Carman made arrangements to have his newest professor awarded an honorary Master of Arts degree by Genesee College, a Methodist Episcopal college based in Lima, New York.[39]

Macoun's luck also improved with regard to field trips. In 1868, George Lawson, now of Dalhousie University, provided the funds to enable the Belleville botanist to accompany an expedition to the Muskokas. The next summer, taking advantage of his longer vacation, he spent the better part of July exploring the north shore of Lake Superior. Most of the specimens collected during this outing went to David A.P. Watt, a Montreal botanist who was also considering a flora of Canada and who had financed the trip.[40] Macoun did, nonetheless, satisfy his desire to collect in some of the lesser-known parts of the province, while adding to his knowledge of those combinations of soil and climate that favoured particular species.

By 1870, John Macoun had come a long way since his days as a young frontier farmer who preferred to puzzle over plants rather than attend to his chores. Now known as 'the professor,' he was a respected member of Belleville's social and intellectual élite and the father of a young family – another daughter, Minnie, had arrived in 1866, and a second son, William, in 1869. Through sheer hard work and determination, he had also emerged as a skilled plant geographer whose field work was regularly financed by other botanists.

Macoun was not satisfied, however. Plants were his passion, and he would never be completely happy until he was able to spend all his days in the field. 'I will do everything in my power towards exploring and thoroughly investigating Canadian Botany but my resources are limited,' he wrote to a British specialist shortly before his appointment to Albert College. 'If I could only raise money enough to pay my expenses and support my family for six months I would do more in that time than I can possibly expect to do during the next two years.'[41] John Macoun knew that the botanical frontiers of the young Dominion, particularly beyond the Great Lakes, were waiting to be explored. And he believed that he was the field naturalist who could best assume this task. All he needed was a chance.

Getting Aboard

John Macoun's involvement in the scientific assessment of western Canada was purely accidental. While collecting specimens in the Owen Sound region in the summer of 1872, he met by coincidence Canadian Pacific Railway engineer-in-chief Sandford Fleming and was invited to take part in the survey of the proposed Yellowhead route for the transcontinental railway. Over the next decade, during five separate exploratory surveys between 1872 and 1881, Macoun examined the agricultural capabilities of various western tracts on the basis of the natural vegetation. Each time, he returned more thoroughly convinced that western conditions were ideal for the large-scale settlement envisaged by Ottawa. Where earlier investigators had warned about summer frosts and insufficient moisture, he spoke of the northward sweep of summer isotherms and rain that followed the plough. Where they had found an irreclaimable desert, he discovered a garden of unlimited potential. Where they had seen a forlorn empty wilderness, he evoked images of 'a land with untold wealth in its soil' where 'life means an unending pleasure.'[1] His enthusiasm knew no limits except the boundaries of the region itself.

These sweeping generalizations, albeit reckless and potentially detrimental in the long run, were precisely what the moment called for. Canada had committed itself to a policy of large-scale colonization in a region that had a chequered history of agricultural experience. The government was anxious that its great hopes for the region be substantiated. By emphasizing only the ideal aspects of the region's character, Macoun provided the scientific justification for this federal policy and, in the process, emerged as the country's leading expert on western Canada. He was rewarded for this work in November 1881. At the conclusion of his regular briefing session with Lindsey Russell, deputy minister of the interior, Macoun was named botanist to what was then titled the Geological and Natural History Survey of Canada. The appointment was immediately confirmed in a short, private interview with the prime minister and minister of the interior, John A. Macdonald. At the age of fifty, John Macoun realized his dream.

At the end of the school year in June 1872, Macoun planned collecting trip to Thunder Bay, this time with the principal of Albert College, Albert Carman. The morning they were to leave, however, Macoun overslept and missed the train. By the time he reached Toronto, Macoun found that Carman had continued on to Sarnia. He therefore elected to proceed to Collingwood and catch a Great Lakes steamer from there.² It was a fortuitous decision, for Macoun soon discovered that one of his fellow passengers on the *Francis Smith* was the renowned railway surveyor and civil engineer Sandford Fleming.

When the Canadian government decided in 1871 to build a transcontinental railway to British Columbia and open the vast prairie interior to agricultural colonization, Fleming had been appointed engineer-in-chief and given the crucial job of determining the most suitable route. He faced a difficult decision, for settlement of the prairie west at that time was anything but certain; it would be a clean-cut experiment. Despite limited agricultural activity in the area for several decades, the record of agricultural experience was more contradictory than conclusive: in some years excellent crops were raised in the virgin prairie soil in the shadow of the fur trade posts or at the Red River Colony; in others the growing crops fell prey to drought, frost and other natural disasters. Many factors critical to the growth of cereal grains – such as the length of the growing period and the amount and seasonal distribution of rainfall in different parts of the region – were also uncertain.³

The issue was further complicated by the existence of two contradictory views of the nature of the western lands – the image of the desert and that of the fertile belt. In the late 1850s, while the potential of the prairies was being debated in London before the Select Committee of Inquiry on the Hudson's Bay Company, the British and Canadian governments dispatched scientific exploring parties to the western interior to gather more reliable information. Although undertaken during a relatively wet period,[4] neither the John Palliser (1857–9) nor the Henry Youle Hind (1857–8) expeditions were impressed with the southern grasslands and instead extolled the merits of the transitional parkland zone or 'fertile belt' along the Assiniboine and North Saskatchewan rivers. This assessment was quite logical, for the party members were unfamiliar with the peculiar plains environment and innocently assumed that the region's treelessness was a sign of aridity. Besides, the expeditions were charged with delineating those areas where agriculture would best be initiated; clearly, the well-watered and wooded prairie parkland was more in keeping with traditional mid-latitude agricultural land. In describing the open prairie region in their reports, however, both Palliser and Hind took the prevailing notion of a great inland desert wholly unfit for agriculture in the American trans-Mississippi west and neatly extended it into British North America to support their pessimistic conclusions about the grasslands' potential. It is quite likely that they would not have been so negative about the plains region had it not been for the perception of a desert south of the forty-ninth parallel. The unfortunate results were the sweeping resource generalizations of a fertile belt and a desert – generalizations that had limited value but would have to be dealt with by future surveys.[5]

This idea of an exceedingly fertile region in the North-West perfectly suited Canadian needs. Believing that Canada faced a future of stagnation and uncertainty unless it was able to break out from the constraints of the Canadian Shield, expansionists had successfully transformed the image of the western interior in the 1850s. Once dismissed in Canadian minds as a frozen wilderness, it was now regarded as a potential agricultural hinterland that would provide the means to empire.[6] This idyllic vision had such a powerful appeal that acquisition of the region became one of the main themes of the confederation movement. In fact by the time Rupert's Land and the Northwestern Territory were transferred to Canada in July 1870, the southern limit of the fertile territory had been extended to the international boundary, even though there had been no further detailed assessment of the North-West since the Palliser and Hind expeditions.[7]

Fleming's job of deciding the best route for the transcontinental railway was thus not an easy one. The Palliser and Hind surveys had done little more than introduce the general concepts of good and bad land, while the Canadian government had secured the region on the assumption that some, perhaps most, of the land was fertile. Before large-scale agricultural settlement could take place, then, a clearer more detailed knowledge of the resources of the North-West was urgently required. It was partly for this reason that the methodical Fleming was headed west in 1872 to acquaint himself personally with the land along Palliser's and Hind's fertile belt – his tentative choice of route for the railroad. This concern with acquiring resource information was also reflected in his decision to invite Macoun to join his party. A man of science himself and one of the founding members of the Canadian Institute in Toronto, Fleming had probably either heard of Macoun and his work or, more likely, had an appreciation of the kind of talents and skills that a botanist could bring to the expedition. He therefore asked Macoun to come along and make a collection of the plants of the North-West, as well as appraise the agricultural prospects of various interior tracts. Although his wife was pregnant with their fifth child, the ever-ambitious Macoun jumped at the opportunity to search for botanical specimens in a relatively unknown region. As he confessed to Reverend George Monro Grant, principal of Queen's University and secretary to the Fleming party, 'This expedition ... is going to give me a lift that will put me at the head of the whole brigade.'[8]

From the outset, Macoun carried out his assignment with infectious enthusiasm. His obsession with collecting specimens is graphically portrayed in *Ocean to Ocean*, Grant's account of the expedition:

The sight of a perpendicular face of rock, either dry or dripping with moisture, drew him like a magnet, and, with yells of triumph, he would summon the others to come and behold the trifle he had lit upon. Scrambling, panting, rubbing their shins against rocks, and half breaking their necks, they trailed painfully after him only to find him on his knees before some 'thing of beauty' that seemed to us little different from what we had passed with indifference thousands of times.[9]

On reaching Thunder Bay, he combined this search for species with a careful inspection of the overall vegetation in an effort to determine the adaptability of the land to cultivation. This procedure was not peculiar to Macoun. Both the Palliser and the Hind surveys had appraised the agricultural prospects of the western interior on the basis of its vegetation. It was also an accepted practice among members of the Geological Survey. Yet because of Macoun's pride in his knowledge of the range and habitat of plant species, his dependence upon such a simple general test was much greater. He never seemed to realize or at least admit that the observation of natural vegetation failed to allow for the possibility that varying and extreme climatic conditions from place to place or from year to year would make agriculture hazardous. Daring assumptions and bold generalizations were the result.

From Thunder Bay, the party worked its way along the land and water links of the Dawson Road. At the North-west Angle, Macoun was assigned his own wagon and driver and was constantly leaping from his seat and running into the woods or meadows to retrieve a specimen that had caught his eye. This behaviour confounded the teamsters, particularly one Scot who sarcastically advised Macoun's driver, 'tell yon man if he want a load o' graiss, no' to fill the buggy noo, an' a'll show him a fine place where we feed the horse.'[10] Undaunted by such ridicule, Macoun kept up his feverish pace in anticipation of what he would find when they reached the open prairie. He was not disappointed. 'When I rose yesterday morning and looked out on the prairie,' he wrote to his wife from Winnipeg, 'I can only say I was astonished for as far as the eye could reach stretched a grassy plain without a fence and nothing to be seen but grass and flowers. In less than an hour I found 32 new plants.'[11]

In Winnipeg, while staying at Government House, the expedition members discussed western Canada's potential with a number of leading citizens, including Dr John Christian Schultz, former leader of the 'Canadian Party' at Red River and now MP for Lisgar, the United States consul, James Wickes Taylor, and Bishop Alexandre Taché, bishop of St Boniface. All praised the qualities of the fertile belt, with the exception of Taché, who maintained that the northern forest region was better suited for cultivation.[12] The party then resumed its march northwestward on 2 August, covering the 900 miles to Edmonton in a record twenty-five days. Here, Fleming decided to break up the survey. Macoun and Charles Horetzky, the expedition's outfitter and photographer, were to be sent north on a reconnaissance survey of the Peace River Pass, while the rest of the expedition continued westward through the Yellowhead Pass to the coast. This additional survey, prompted by a recently published pamphlet advocating a northern route for the railway,[13] was characteristic of Fleming's thoroughness – a thoroughness that eventually translated into ten years of surveys at a cost of $4,166,187. Macoun interpreted the assignment as a chance for further glory. In a letter explaining the change in plans to his wife, he disclosed that, 'Mr. Fleming told me before leaving ... to do it up thoroughly and honestly and all the credit would be mine.'[14]

Although he tried to learn as much as possible about the Peace River country during his brief stay in Edmonton, Macoun had little idea of what lay ahead when he and Horetzky set off on 6 September. Six weeks later, however, he was lauding the country as being ahead of anything he had seen in terms of beauty and fertility. 'I would prefer risking wheat on any part of the prairie passed over today,' he recorded in his notebook on 14 October along the trail between Dunvegan and Fort St John, 'than in the neighbourhood ... of Edmonton. Nothing in either soil, plants or climate would cause me to hesitate in giving this opinion.'[15] Such an enthusiastic endorsement of the region was not surprising under the circumstances. During the first leg of their trek between Edmonton and Lesser Slave Lake, Macoun and Horetzky had struggled amid almost constant rain and wind through a broken swampy land. But as they drew near the mighty Peace, a few miles below the mouth of the Smoky River, the trail became steadily less dreary, and they soon found themselves basking under sunny skies in a lush prairie parkland. This dramatic change in travelling conditions was in itself enough to create a favourable impression. Macoun, however, probably did not expect to encounter such a rich vegetation at so high a latitute so late in the season. He consequently developed an exaggerated appreciation of the land's potential.[16] The Professor, moreover, had had scant opportunity to examine the land between Winnipeg and Edmonton – Fleming had been determined to average at least forty miles a day – and what Macoun had seen of the fertile belt had not overawed him.

Despite these encouraging findings, relations between Horetzky and Macoun steadily deteriorated to the point where the pair were close to blows. Earlier at Dunvegan, the seasoned Horetzky had, in the

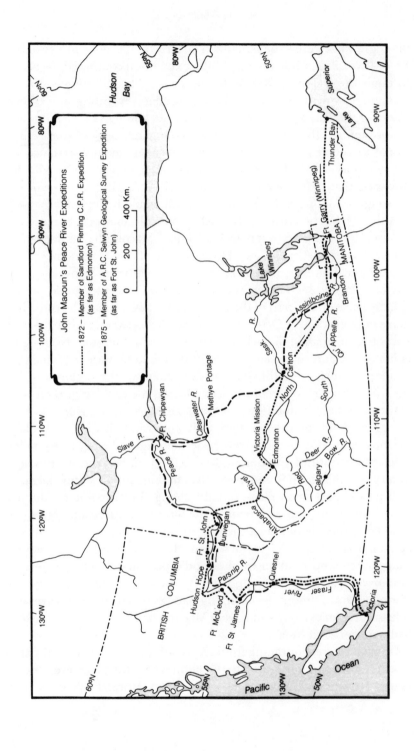

John Macoun's Peace River Expeditions

1872 – Member of Sandford Fleming C.P.R. Expedition
(as far as Edmonton)

1875 – Member of A.R.C. Selwyn Geological Survey Expedition
(as far as Fort St. John)

0   200   400 Km.

Hudson Bay

Lake Superior

Thunder Bay

Ft. Garry (Winnipeg)

MANITOBA

Brandon

Assiniboine

Qu'Appelle R.

Lake Winnipeg

Sask. R.

North Sask.

Carlton

Victoria Mission

Edmonton

South Sask.

Red Deer R.

Bow R.

Calgary

Ft. Chipewyan

Clearwater R.

Methye Portage

Slave R.

Peace R.

Athabasca River

Dunvegan

Ft. St. John

BRITISH COLUMBIA

Hudson Hope

Parsnip R.

Ft. McLeod

Ft. St. James

Quesnel

Fraser River

Victoria

Pacific Ocean

Professor's words, 'begun to show the cloven hoof,'[17] trying unsuccessfully to convince Macoun to return to Edmonton because of the lateness of the season. Now, at Fort St John, in northeastern British Columbia, the former Hudson's Bay Company clerk wanted to ascend the north Pine River alone in search of a previously unknown pass reported by the local Indians. The botanist, however, remained determined to push on even if it meant leaving his bones in the mountains. The two men consequently stayed together as far as Fort St James, which they reached on 14 November. Macoun then fled south over the snow-laden Blackwater Trail and Cariboo Road to Victoria,[18] while Horetzky explored westward along the Skeena River to the Pacific coast.

By the time Macoun finally returned home to Belleville in January 1873, he found that his wife had given birth two months earlier to their fifth and final child, Nellie. The professor had felt extremely guilty about his unexpected, prolonged absence in the North-West and had repeatedly confirmed his love for his wife and young family in his letters home. 'When I awake in the morning my first thoughts are of you and the little ones,' he had written from Winnipeg. 'Think of me at the same time and I shall feel that I am not alone in the world but there is one heart that turns to me in loving remembrance.'[19] Such outpourings undoubtedly eased Ellen Macoun's concerns. She was not happy with her fate as a summer widow, especially when the children were so young. At the same time, she had apparently realized from the beginning of their marriage that he loved plants almost as much as he loved her and that it would be foolish to attempt to keep him from the field. She consequently had little choice but to accept her husband's absences stoically, while relying on hired help and close family friends to fill the void. As for the children, they knew that their father's field trips were a fact of life and that his return would bring little presents, as well as stories of his latest adventures. In the meantime, they were expected to obey their mother and help out at home in whatever way they could. A strong sense of family and devotion to their parents consequently developed; though in keeping with their father's own upbringing, the children were also encouraged to develop an independent cast of mind and to be as self-reliant as possible.

Upon resuming his teaching duties at Albert College, Macoun prepared at Fleming's request a summary account of his activities for the 1874 railway report. In reviewing his 3,000-mile journey from Winnipeg to Edmonton and then up the Peace River through the mountains and down to the coast, he found that his collections indicated a regional uniformity in the vegetation of the western interior. 'The hill-top, the plain, the marsh, the aspen copse, the willow thicket,' he reported, 'each had its own flora throughout the region, never varying and scarcely ever becoming intermixed.'[20] These findings, when combined with a careful rereading of Humboldt's *Cosmos*, suggested that the western interior was a complete geographical entity – 'there is a great uniformity respecting soil, humidity and temperature throughout this region.'[21] They also convinced him that mean summer temperature as opposed to mean annual temperature was the more important statistic. With these ideas in mind, Macoun put together his report, the majority of which was devoted to his activities after leaving Fleming in Edmonton. In his brief review of his survey in the North Saskatchewan country, he expressed concern over the seeming prevalence of summer frosts and concluded 'there is no better country for raising stock.'[22] He then went on to describe in considerable detail the rich soil, rank vegetation, and warm climate of the Peace River country, arguing that the region was not simply comparable to the Edmonton area but possessed the greatest agricultural potential of any district that he had observed.

When Dr A.R.C. Selwyn, director of the Geological Survey of Canada, read the proof sheets of Macoun's report during a visit to Ottawa in April 1874, he asked the professor whether he would accompany him to the Upper Peace River country the following summer. This invitation represented a complete reversal in Selwyn's attitude towards botanical investigation. Just a few weeks before Macoun had been engaged by Fleming to accompany his expedition, James Brown, the Conservative MP for Hastings West, had written to the under-secretary of state for the provinces, recommending the botanist's appointment to the Survey. The letter was forwarded to Selwyn who tersely replied, 'No useful purpose can be served by connecting it with the Geological Survey; the objects and requirements of the two are entirely distinct, and I could not be responsible for botanical work of which I have no knowledge or experience.'[23]

Brown's request was not new. The question had been raised as early as 1858, when the *Canadian Journal*, in response to the publication of the natural history reports of the Pacific Railroad Survey, asked, 'why is there not a Zoologist and Botanist attached to the Geological Survey? ... Surely the expense of a naturalist with a separate canoe which he would require, would not have added considerably to the costs of the Survey, and we should have had now such valuable materials for the elucidation of the flora and fauna of the country as can seldom be obtained.'[24] Selwyn's predecessor, Sir William Logan, appreciated the logic of this argument and had occasionally employed a natural history collector during some of his surveys. He was not willing to go so far as to establish a permanent position, however, because the Survey was not responsible for this kind of work, and hence the agency's funds could not be used for this purpose. He had also never been completely satisfied with the results of these investigations.[25]

Selwyn was in a similar position. Since assuming the directorship in 1869, he had dearly wanted to make the Survey as wide a service as possible. 'I am aiming to make the work of our survey thoroughly reliable,' he told a good friend, 'and that the facts stated and the conclusions arrived at shall be the results of not what may have been said or written by others now or in the past, but in all cases of renewed and careful observations within the areas reported on.'[26] Yet the Survey's financial and manpower resources were entirely out of line with its responsibilities. Created in 1842 by the United Province of Canada for the express purpose of investigating and aiding in the development of the mineral resources of the colony, the Survey had from the beginning faced the problem that many of the outlying regions needed general exploration and topographical mapping as well as detailed survey work. This workload was increased significantly in 1870 and 1871 when the Survey was given the major responsibility for mapping and reporting on the geology of the newly acquired Rupert's Land and British Columbia. Given the expanse of territory assigned to it, then, the Survey could do little more than concentrate on description and collection. There were simply no funds for special botanical field parties, not even for an extra canoe.

The other major stumbling block was the Geological Survey's responsibilities. The government agency had been founded in the belief that scientific knowledge was to be used for practical and immediate results; it did not believe in undertaking scientific research for its own sake.[27] Dr Selwyn realized that this philosophy was particularly important to Canadian legislators, who approved the Survey's annual expenditure, and he could not see that botany would meet this requirement. On the contrary, it seemed to have little direct relationship to the Survey's geological duties. He therefore saw little alternative but to inform the under-secretary of state for the provinces that, 'I do not think that such an appointment is desirable in connection with the Geological Survey.'[28]

Macoun's report on his activities during the Fleming expedition and the conclusions he was able to reach on the basis of his field work removed Selwyn's reservations about the appropriateness of having the Survey conduct a botanical investigation. The Professor had made botany relevant to one of the national tasks at hand – large-scale settlement of the North-West – by demonstrating how it could provide practical knowledge of the potential agricultural resources of the region. He had discovered plant species flourishing around Edmonton and along the Peace River that were also found growing in the farming districts of Ontario. The potential applications of Macoun's work were thus too great to be ignored. Selwyn not only asked him to take part in his forthcoming expedition to the Upper Peace River region, but also spoke in terms of a permanent appointment as botanist. In order to make Macoun's relocation to Survey headquarters in Montreal as attractive as possible, he also made arrangements with Sir William Dawson for the Professor to work part-time at McGill.[29]

Macoun was ecstatic about Selwyn's offer. Writing from Edmonton during a bout of homesickness in August 1872, he had promised his wife that the Fleming expedition would 'cure me of rambling ... you can calculate on my staying.'[30] Back in Belleville, however, Macoun found that the letters he had written during the expedition to the *Belleville Intelligencer* and the *Canadian Christian Advocate* had made him a local celebrity, and he was fêted at Albert College with the traditional oyster supper. He had also had ample time to reflect upon the journey and the opportunities that future expeditions held for his botanical studies in particular and, more importantly, his career in general. In fact, the time he had spent with Fleming and Grant had probably instilled in him a sense of the great promise that the North-West held for the young Dominion and the role he could play in that destiny. The idea of being named Survey botanist consequently had great appeal. 'I sincerely hope you will be able to put in "the word in season,"' he wrote to Fleming about Selwyn's proposition, 'and that through your instrumentality the dream of my life will be fulfilled.'[31]

Securing Macoun a permanent position on the Survey staff, however, was not an easy matter. The badly slumping Canadian economy and the inability to secure a private builder for the transcontinental railway because of the 'Pacific Scandal' caused the new Liberal government of Alexander Mackenzie to adopt a more sober policy on westward expansion: the line would be built as a public works project, as settlement and the country's ability to pay warranted. With the completion of the railroad no longer urgent, Selwyn was unable to convince the minister of the interior, David Laird, to hire the botanist on a full-time basis; it also appeared uncertain whether funds would be made available for Macoun to accompany the Peace River expedition. All the director could suggest to Macoun was that he 'use whatever influence you may have with the Members of the Cabinet, in getting them to support the proposition when it comes before them for consideration.'[32]

A Conservative in politics, Macoun greatly doubted his chances of employment and made plans to return to the Lake Superior region in the summer of 1875, intending to sell specimens to cover his expenses.[33] But in March 1875, more than a year after being first contacted by Selwyn and only a month before he was to leave, his appointment as botanist to the Selwyn expedition was confirmed. Ironically, Charles Horetzky, Macoun's nemesis from the Fleming survey, may have been a factor in this decision. When Horetzky reported back to Ottawa in March 1873, he too had been highly enthusiastic about the Peace River country's potential and had suggested routing the railway through the unsurveyed Pine River Pass and then along the Skeena River to Port Simpson. Fleming, however, remained committed to the Yellowhead route and, when he learned that Horetzky was privately trying to win over Alexander Mackenzie, then Liberal opposition leader, he dismissed the surveyor. Horetzky retaliated by publishing a series of articles in the Ottawa *Citizen* in October 1873 in which he advocated a more northerly route for the railway. He also intensified his lobbying efforts, dedicating to Mackenzie, now prime minister, his account of the 1872 trip through the Peace River Pass to the coast, *Canada on the Pacific*. In the end, the new Liberal administration adopted Fleming's choice of route for the railroad.

Horetzky's persistence nonetheless paid off, for he was rehired at Mackenzie's urging as an exploring engineer.[34] Through Horetzky, too, Macoun's work may have come to the attention of the prime minister and resulted in his appointment to the Peace River expedition.

Responsible for gathering information on the region's flora, climate, and agricultural potential, Macoun saw the assignment as a chance to solidify his claim to the position of Survey botanist. 'I believe now as I always did that my absence now was a necessity,' he reassured his wife in a letter as he headed westward on the Union Pacific Railway to San Francisco in April 1875. 'I assure you,' he continued, 'that neither honour nor ambition shall tempt me to desert you again.'[35] In the meantime, he was leaving nothing to chance. In an attempt to publicize his work during the expedition, he had made arrangements with the London *Advertiser* to provide regular instalments describing his western exploits at five dollars per letter. He also secretly planned to collect multiple duplicates of the rarer plant species that he would later sell to specialists, thereby enhancing his reputation as a formidable collector. Finally, he carried a letter of introduction from George Barnston, a former Hudson's Bay Company factor and fellow botanist, in order to secure special treatment at the various fur trade posts along the Peace.

The Selwyn expedition was essentially a retracing in reverse of the route followed by Macoun three years earlier. From Quesnel on the Fraser River, the party headed overland by pack train to Fort McLeod in the northern British Columbia wilderness and then, continuing by water, descended the Peace River to Fort St John. Throughout this period Macoun went about his field work with unwavering militancy. Each day on the trail, he relentlessly scoured the landscape, placing multiple specimens of every plant that he came across in his wicker collecting basket. Whenever the party stopped, he labelled the specimens with a rough geographical location and date and then placed them on herbarium sheets, covered them with felt wrappers, and tightly strapped them between two boards. Over the next few days, the wrappers in the plant press would be regularly replaced until the specimens were completely dried. They were then bundled and stored in wooden crates, not to be examined again until his return home. Some 20,000 specimens would eventually be collected in this manner that summer.[36]

At Fort St John, while Selwyn made a second futile attempt to locate the Pine River Pass, Macoun temporarily left the expedition to continue his floral studies a few hundred miles downriver with a Hudson's Bay Company trader, W.F. King, who was going to meet the annual supply boats from Fort Chipewyan. Setting off on 4 August in a small cottonwood dugout canoe, the pair made good progress and within a week had reached Fort Vermilion. Although warned that the supply brigade that was supposedly already on its way upriver had been delayed by at least two weeks, the pair pushed on. Foolishly battling strong headwinds that whipped up the ever-widening river valley, they soon lost their outfit, except for Macoun's specimen box, when the canoe capsized in a cascade well over two hundred miles short of their goal. It was only with the utmost exertion that the men, hungry, exhausted, and somewhat delirious, finally reached Chipewyan on Lake Athabasca on 23 August.[37]

Macoun's arrival at the famous fur emporium of the north coincided with the annual rendezvous of traders from the various outlying districts. These men, in particular Roderick MacFarlane, chief factor of the Athabasca district, apparently saw in the Professor a chance to generate greater government interest in the region and bring about much-needed improvement in transportation and communication.[38] They therefore welcomed Macoun into their counsels and, during the course of their deliberations over the next few days, assured him that summer frosts were negligible and that the whole country, including the region north of sixty degrees latitude, was fit for the cultivation of wheat and barley. Macoun welcomed such information, for it confirmed his own findings that summer. During the descent of the Peace he had carefully inspected the gardens and cultivated fields at each of the settlements along the way, including those of Chipewyan and the nearby Catholic mission. Repeated comments that it had been a particularly dry summer served only to enhance his astonishment that the crops of vegetables and grains should be so bountiful and the natural vegetation so luxuriant. Searching for a possible explanation, he reasoned that the soil had to be rich to support such growth and that summer temperatures had to be high for the crops to mature so early.[39]

The traders' information suggested that this thinking was not only sound but that it applied to most of the region. Indeed, their reports, when combined with Macoun's discoveries and his own sense of infallibility, caused him to experience a prophetic vision at Chipewyan. He believed that he, and he alone, had discovered the true character of the region where all previous explorers had failed and that it was only a matter of time before the accuracy of his assessment would be borne out:

Writing here at Chipewyan in the centre of the Wild North Land ... the vastness seems to overpower the mind and cause that benumbing feeling which we are prone to feel when in presence of something we cannot grasp ... the soil wherever tried throughout the whole extent of this vast region gives enormous returns for little labor, giving promise of the day when the land will be filled with a busy multitude who instead of living by the chase will cultivate the rich soil and develop the unbounded resources of this wonderful land.[40]

This appreciation of the region's destiny, in turn, dictated a new role for the botanist. His work on the Selwyn expedition was no longer simply a means to a permanent position on the Geological Survey staff. He henceforth saw himself as the advocate for the great resource potential with which the Creator had endowed his adopted country. He concluded: 'Sitting in my room and looking over the immensity I have passed over and thinking of the distance yet before me I am led to exclaim What a wonderful heritage has fallen to us! O that our statesmen would rise to the importance of the trust committed to them! My duty becomes plainer as I consider the matter and I shall not rest until the Canadian Public knows the value of this immense country both as regarding its resources and its capability of development.'[41] Macoun would tackle this self-assigned duty with the same enthusiasm, drive, and energy that characterized his field work. From this point forward, he had a great faith in the western country's destiny and an even greater one in his role in revealing that destiny.

Having been advised by the traders not to try to rejoin Selwyn because of the lateness of the season, Macoun departed for Winnipeg via Methye Portage and Fort Carlton on 2 September. He left Chipewyan armed with sheaves of wheat and barley that he had secured from the nearby Catholic mission. These samples 'created quite a sensation at Winnipeg'[42] and would later win a silver medal at the 1876 Centennial Exhibition in Philadelphia. He also did not forget his pledge. Whereas his early journal entries had been characterized by notes on the vegetation, climate, and general travelling conditions, his remarks from Chipewyan forward now contained the occasional prophetic passage. 'Long after the noises ceased,' he wrote at his campsite near the present day Athabasca oil sands, 'I lay and thought of the not distant future when other sounds than these would wake up the silent forest and the white man with his ready instrument steam would be raising the untold wealth which lies hidden underneath the surface. It has never entered into the brains of our most enthusiastic citizen or statesman the wealth that lies hid in this land.'[43] He made a similar entry southeast of the Touchwood Hills as he worked his way towards Winnipeg: 'But is this solitude to last – No! 200 miles eastward a low steady tramp is heard – it is the advance guard of the teeming millions who will yet possess this land ...'[44]

Macoun reached Belleville on 13 November 1875 and immediately started work on a preliminary sketch of his findings. Before completing it, he travelled to Montreal in early December to consult with Dr Selwyn, who himself had just returned from the west coast, about his status with the Survey. Selwyn knew that the agency's limited appropriation prevented him from keeping the botanist on salary but chose not to tell him during his visit; instead, he advised Macoun that he too would be reporting favourably on the Peace River country and asked the Professor to prepare a display collection of Canadian plants for the Philadelphia Centennial Exhibition. It was only after Macoun returned home that Selwyn told him his salary could not be continued beyond 31 December.[45]

The letter came as a great blow to Macoun, particularly since Selwyn had given no indication of his true intentions during their Montreal meeting. True to character, however, Macoun soldiered on, confident that the plant display for the exhibition would provide another opportunity to secure international recognition and a possible government position. He had also not seen Prime Minister Mackenzie, as he had promised himself during his week-long stay at Chipewyan. 'All those things I noted,' he confidently wrote to MacFarlane on 19 December 1875, 'will be laid before the head of Government and I know some of them will be acted upon.'[46] The Professor got his chance early in the new year when he went to Ottawa to report to the minister of the interior, David Laird, who in turn informed him that the PM wanted to see him. Macoun held nothing back, telling Mackenzie among other things that it was possible to travel two hundred miles in the North-West without seeing an acre of bad land. At the end, an incredulous Mackenzie remarked 'I canna believe it,'[47] and remained committed to sending the railway along the proposed Yellowhead route.

Returning to Belleville, Macoun turned to his unfinished report of his 1875 field activities, determined to demonstrate beyond any doubt the great worth of the Peace River country and, thereby, the accuracy of his assessment. In putting together his report, Macoun could simply have drawn on his own observations and the experience of the traders. Mackenzie's rebuff had shown, however, that firsthand information by itself was not enough. It had to be accompanied with an explanation for these phenomenal growing conditions. The Professor consequently began to look for possible theories in much the same manner that he looked for new species in the field. One idea that seemed particularly appropriate to his findings was the belief that yields were greater near the northern limit of successful growth because plants matured more slowly. This idea was a popular botanical law in the 1860s[48] and had been raised by the American consul, James Taylor, in Winnipeg when he saw the fine samples of wheat and barley that Macoun had obtained near Chipewyan. The other theory that appeared to fit the Professor's discoveries was the work of American climatologist Lorin Blodget. In his provocative book *The Climatology of North America* (1857), Blodget had combined limited meteorological data with bold generalization to produce summer isothermal lines that extended northward into the western interior and seemed to indicate a favourable temperature for agriculture. Blodget's isotherms had been used almost two decades earlier by Canadian expansionists pushing for acquisition of the North-West.[49] Macoun now embraced them as the explanation for the Peace River's exceptional growing conditions, despite the fact that temperature extremes, especially early frost, had played havoc with western agriculture since its fur trade beginnings. As Archbishop Taché, a long-time resident of the region, warned, 'experience proves that they [isothermal lines] are not to be depended upon. These lines are fundamentally wrong, for, I repeat ... a single night is sufficient to destroy all analogy with the climate of the country to which they refer.'[50] Macoun, however, was more concerned with emphasizing the warmth of the growing season than with measuring the actual length and, from the nature of the western vegetation, was convinced that frosts would never do any serious damage.

Macoun's eventual report was a cleverly crafted endorsement of the settlement capabilities of the Peace River region. 'It would be folly to attempt to depict the country,' he warned, 'as it was so much beyond what I ever saw that I dare hardly make use of truthful words to portray it.'[51] Dare he did, however, not only in his report but in lectures before the Ottawa Literary and Scientific Society and the Montreal Natural History Society. These entertaining talks made Macoun something of a public figure and, in his own words, 'I began to consider myself of more importance than I had before and took my place in the city [Belleville] accordingly.'[52] This growing popularity as an expert on the North-West also had its frustrating side. Here he was, the discoverer and promoter of the northland's great potential, teaching in Belleville and not on the Survey staff because of the Liberal administration's indifference to his work. 'I would like to stand before the assembled wisdom,' an exasperated Macoun unburdened himself on 11 March 1876 to Malcolm McLeod, the Ottawa pamphleteer who first advocated the Peace River route for the railway, 'and tell them what I think of them and of the great country they do not even know by name. I am sick of parties but more particularly of those who govern now.'[53] The Professor was given the chance to make his opinions known only two weeks later when he was called before the House of Commons Select Standing Committee on Agriculture and Colonization.

Macoun appeared before the committee for six hours on 25 and 26 March 1876. He outlined the various resources of the Peace River district, recounted his discussions with the traders at Chipewyan and, when asked whether a railway should be sent through the region, remarked, 'It is the garden of the Dominion.'[54] The committee was also interested in knowing his views on the southern prairie district and questioned him several times during the course of his testimony about its agricultural prospects. Although Macoun had never personally examined the region and repeatedly said so, he did not shrink from giving an opinion. He suggested that the Canadian portion of the American desert had a better climate and was suited for stock raising and possibly wheat cultivation. At one point, he even responded, 'All accounts agree ... in saying it is the garden of the country.'[55] Both the committee's questions and Macoun's answers about the prairie grasslands were significant in that they signalled the beginnings of a scientific reappraisal of the area that Palliser and Hind had earlier condemned. This re-examination of the South Saskatchewan country had been initiated by the recent work of George Mercer Dawson, a geologist attached to the North American Boundary Commission Survey. During the 1873 and 1874 field seasons, Dawson had explored the country along the forty-ninth parallel from Lake of the Woods to the Rocky Mountains. His subsequent report, which did much to foster his reputation as one of Canada's foremost geologists, included a judicious assessment of the potential of the southern region. He stated that the second and third prairie steppes, with their light soils, scanty precipitation, lack of wood, and high incidence of frosts, did not compare favourably with the fertile belt. At the same time, he took great care to point out that 'this country, formerly considered almost absolutely desert, is not – with the exception of a limited area – of this character; that a portion of it may be of future importance agricultural-ly, and that a great area is well suited for pastoral occupation and stock farming.'[56] In fact, the area available for cultivation purposes could probably be enlarged, according to Dawson, by putting an end to prairie fires and planting trees or ploughing the ground.[57]

George Dawson's Boundary Commission report was not designed to divert Canadian energies away from the North Saskatchewan country to the southern plains. He made it quite clear that 'the fertile belt ... must form the basis for settlement' and that 'vast areas of the western plains ... must remain as pasture grounds.'[58] The geologist was equally determined, however, to correct the false impressions about the prairie district that had been generated by the Palliser and Hind expeditions. Whereas the earlier surveys had projected the Great American Desert into Canadian territory, he observed that a large part of the American plains had been found not to be typical desert at all. Unlike the earlier explorers who imposed their values on the grasslands and so naturally concluded that the region was deficient, Dawson argued that the progress of settlement should be 'a natural growth taking advantage of the capabilities of the region.'[59] In sum, it was an exceptional report that could have laid the basis for the sound development of the plains district. But in Macoun's hand, it provided the opening for a complete reassessment of the region's potential that went far beyond what Dawson had ever imagined.[60]

Macoun was quite familiar with Dawson's work for the North American Boundary Commission Survey. During the winter of 1874–5, perhaps at the suggestion of Selwyn or Fleming, he had determined as best he could the grasses and carices that the geologist had collected along the forty-ninth parallel. Later, during his Montreal meeting with Selwyn in December 1875, he had examined the remainder of Dawson's collections at Survey headquarters. When Macoun subsequently told the geologist that he had looked over his herbarium during his absence, Dawson penned on the back of the letter: 'an enthusiastic *collector* of such beware!!'[61] It was not his plants, however, that Dawson had to worry about.

During his questioning before the Commons Committee on Agriculture and Colonization, Macoun mentioned that he was aware of Dawson's conclusions but did not discuss them at any length. It is quite likely that Macoun had not had the opportunity to mull over their ramifications, since he was busy with his own report at this time, as well as with the plant display for the Philadelphia Centennial Exhibition. During the early summer of 1876, however, Prime Minister Mackenzie, who also doubled as his own minister of public works, seemingly overcame his earlier scepticism about the botanist's reliability and asked Macoun to prepare an assessment of the agricultural worth of the entire western region for the January 1877 Pacific railway report. It is not clear why such a report was commissioned at this time nor why Macoun was given the task. Sandford Fleming left for England on an extended leave of absence in July 1876 and might have wanted, prior to his departure to pull together all the information that the various railway surveys had gathered. The Liberal government may also have been acting on the suggestion of Dawson himself, who observed in his report: 'Accurate and detailed information ... is now in process of accumulation over a great part of the North-West, and it will ere long be possible to estimate the probable value of the whole interior portion of the Dominion.'[62] Whatever the reason, Macoun was a logical choice for the assignment – an assignment that presented him with an ideal opportunity to draw on Dawson's work to further his image as one of the few men who understood the real potential of the Canadian North-West.

The Professor entertained no doubts about the accuracy of Dawson's assessment of the prairie region. Besides respecting the geologist as an accomplished scientific investigator, he probably assumed that while he had discovered the true character of the Peace River country, Dawson had done the same, or at least was providing the clues, for the grasslands district. In outlining the boundaries and characteristics of the various regions of the North-West, then, Macoun drew quite heavily on Dawson's report. In fact, just as he appropriated theories that suited his findings, he carefully marshalled all the favourable information that could be culled from previous explorers' reports. What emerged out of this 'cut-and-paste' exercise was a highly encouraging portrait of Canada's western domain, including the semi-arid third prairie steppe. Macoun did mention those areas where land was particularly poor, but in general intimated –without stating outright – that Palliser and Hind had been wrong in their evaluations. Not to be outdone by the earlier explorers, he also offered a generalized resource summary of the land available for settlement:

If a line be drawn from the Boundary Line where it is intersected by the 95th meridian in a north-westerly parallel, we shall have the base of an isosceles triangle, which has its apex on the 115th meridian, where it intersects the 49th parallel, one side being the Boundary Line and the other the Rocky Mountains. This triangle encloses at least 300,000 square miles or over 200,000,000 acres of land.[63]

Of these 200 million acres, he roughly classed 80 million as arable land and 120 million as pastures, lakes, and swamps. In particular, 200,000 of an estimated 5,120,000 acres of 'dry arid pastures' in the South Saskatchewan country were suitable for grain cultivation; the remainder, far from being sterile, could be used for grazing purposes.[64]

Given Macoun's belief that he was lifting the veil that had shrouded the North-West, these figures were not really surprising. But Dawson had also warned that to be successful development would have to adapt to the peculiarities of the region. The failure to recognize the treeless prairie as a distinctive North American physical environment had accounted, in part, for Palliser's and Hind's pessimistic conclusions. And even though he was reaching opposite conclusions about the potential of the grasslands, Macoun was equally guilty of imposing unsubstantiated suppositions on the region. In 1872 and 1875, he had assessed the capabilities of the Peace River country on its ability to support plant species that grew where agriculture was already being practised in Ontario. Similar comparisons between the flora and the summer temperatures of Ontario and western Canada were now made in his report's conclusion. He also noted that much of the prairies had once been covered by trees and that forested land, according to his Ontario farming experience, translated into adequate precipitation and cultivated fields. 'None of the prairie country, except that south of the Missouri Couteau [third prairie steppe] is naturally so deficient in rainfall to prevent forest growth,' he declared. 'It is to be doubted that any deficiency exists.'[65]

Prime Minister Mackenzie was not pleased with Macoun's enormous estimates. He had cautioned the Professor not to draw on his imagination and, upon receipt of the report, dismissed him as a 'cracked-brained enthusiast.'[66] A permanent position with the government, however, still appeared a possibility. In early 1877, David Mills, the new minister of the interior, tabled changes to the Geological Survey Act. Up to then, the organization was essentially a Crown agency whose mandate was renewed every five years by Parliament. Mills now designated the Survey a permanent branch of the Department of the Interior, dependent on annual parliamentary grants. He also gave it a new title, 'The Geological and Natural History Survey of Canada,' that carried with it new duties: the agency would now be responsible for examining the flora and fauna of the Dominion, in addition to its regular geological duties. Any natural history items collected were to be displayed in the Survey museum, which at that time housed only geological and mineralogical specimens.[67]

The new Survey responsibilities brought Mills under attack during the second reading of the bill in February 1877. Sir John A. Macdonald, leader of the opposition, was afraid that the scheme was too large in its scope. 'The Survey was instituted,' he scolded the minister, 'as its name indicated, for the purpose of making a geological survey of the whole country and for other inquiries in the nature of physical sciences.'[68] Mills met this criticism head on. 'Questions concerning flora and fauna were also intimately connected with economic geology and the agricultural capabilities of the country,' he retorted, 'and it was of the utmost importance, therefore, if the geological investigations were to be of any economic value that those matters should be carefully considered.'[69] Macoun could not have asked for a better defender. The minister of the interior was arguing for exactly the kind of work that Macoun had performed during his surveys of the North-West in 1872 and 1875 and seemed to be preparing the ground to hire someone with the botanist's talents. But there was a major obstacle. Despite the additional duties, the bill did not increase the Survey's annual appropriation, and consequently the Professor was not hired.

Macoun plodded on. Because geologists were now required to collect biological specimens as part of their regular field duties, Dr Selwyn regularly turned to him for help in their classification.[70] The Professor undertook this voluntary work largely in the interests of his floral studies, as well as to advance his employment prospects when and if additional funds were made available for Survey operations. The real turning point in his fortunes, however, occurred when the Macdonald Conservatives under their 'National Policy' banner romped back into office in the 1878 fall federal election. The new government proposed to link a tariff-protected eastern manufacturing base with an agriculturally oriented western hinterland by means of an all-Canadian railway. The construction of the Pacific railway therefore acquired a new urgency – it was fundamental to the settlement of western Canada on which the success of Macdonald's plan of economic nationalism depended.

During their five years in opposition, the Conservatives had changed their position on the route for the transcontinental railway and had begun to advocate a more northerly pass through the Peace River district. Fleming, on the other hand, had never wavered from his initial decision and, in his last report before the election, continued to argue that the rail line would best serve the interests of the country if it followed his Yellowhead route. The new Macdonald administration was in a potentially embarrassing position. It could not simply reverse its stand without giving the impression that the former Liberal government had been right all along. Sir Charles Tupper, the new minister of railways and canals, consequently decided that the government had little option but to have the merits of the two routes investigated one more time, and he began assembling survey parties for the 1879 field season.[71]

Macoun was approached by his old Belleville friend Mackenzie Bowell, now minister of customs, about heading one of the parties. He was naturally elated but told Bowell that he wanted something more permanent. The minister promised to pursue the question with Tupper and was soon able to report back that an informal arrangement had been worked out whereby Macoun would serve as explorer for the Canadian government in the North-West Territories as long as the Conservatives remained in power.[72] Although the exact nature of Bowell's negotiation efforts are not known, the Professor's unprecedented appointment raises some important questions. In holding out for a permanent position, was Macoun simply confident that Tupper required his expertise? Or had he led Tupper to believe that his field work would result in promising findings? Similarly, was Tupper hiring the botanist on his own merits or did he demand certain results in return for the position? Clearly, both men had something to gain through the appointment.

Macoun's first assignment in his new role was an exploration of the prairie district south of the Carlton trail between Winnipeg and Edmonton and north of the fifty-first parallel. The only person to explore the southern grasslands that season, he had evidently been sent by Tupper to refute Palliser's and Hind's statements about the district's barrenness. Within a week of Macoun's departure for the field, the minister of railways began to emphasize the overall fertility of the lands of the western interior. 'We believe that, today, ... we have vast regions only partially explored which are not second to any lands in the West,' he said in the House of Commons on 10 May. 'We believe that we have there the garden of the world.'[73]

Such extravagant claims were necessary because of the Conservative railway policy. The government hoped to attract a private builder for the Pacific railway by the offer of a large western land grant, and the prospect of wonderfully fertile lands would greatly facilitate this scheme. Yet apart from political considerations, Tupper's speech reflected the extent to which a large part of the North-West was perceived by Canadians as a possible agricultural garden. An incredible volume of popular literature during the 1870s waxed enthusiastic about the resources and capabilities of the region, while downplaying the extent to which the Great American Desert extended into Canada. Thomas Spence, clerk of the Manitoba Legislature, observed that 'the great portion of this section of territory [Saskatchewan country] ... is as rich in soil as any part of America, and presents the natural advantage of being ready for the plough without the trouble of clearing.'[74] James Trow, chairman of the Commons Committee on Agriculture and Colonization during Macoun's 1876 appearance, shared these sentiments: 'On the prairie, nature has prepared the soil for immediate use, all that is required being to fence the fields and begin ploughing.'[75] Even the prudent Fleming reported before the Royal Colonial Institute in London in 1878 'that a great breadth of the country previously considered valueless may be used for pastoral purposes and some of it ultimately brought under cultivation.'[76] No writer, however, ever went so far as to suggest that an area of bad land did not exist or that the western lands were all equally good.[77]

Macoun, for his part, was happy to be involved in official field work again. Since his return from the Selwyn expedition, he had become disillusioned with government and pessimistic about his chances of an appointment to the Survey staff. He had more or less been forced to accept the fact that his Conservative sympathies were working against him and that the Mackenzie Liberals could not be expected to set aside the understood rules of patronage and create a new position for him. It was a time when politics as much as ability, whether scientific or otherwise, determined appointments to many civil service positions. A new Conservative administration committed to building the railway and settling the west as quickly as possible, together with a quasi-permanent position as government explorer, now made all the difference. After waiting almost four long frustrating years for this opportunity, he was determined to make the most of it.

The Macoun who set off in the late spring of 1879, however, was more than the self-assured tireless collector of the previous two expeditions. He was keenly aware of the bright and prosperous future that Canada foresaw for the region – that the western interior had come to be regarded as an agricultural Eden where the best features of British civilization were to be recreated. The Professor, in fact, shared this vision, thanks largely to his involvement in the debates over the best route for the railway and the fertility of the various western regions. 'Want, either present or future, is not to be feared,' he would later write, 'and man living in a healthy and soul invigorating atmosphere will attain his highest development, and a nation will yet arise on these great plains that will have no superior on the American continent.'[78]

Macoun's 1879 assignment was in essence, then, a mission to substantiate these great hopes of what the west ought to be. He was not, however, a charlatan who, in the words of his most vociferous critic, 'offended in the light of knowledge, history and experience'[79] to suit the needs of the period. His great enthusiasm for the region was matched by a great faith in his scientific abilities – the one fed the other. He genuinely believed that he was right about the North-West's destiny and that, in time, as he was always in the habit of remarking, his statements would be shown to be the truth. It appeared that this process had already started. While Macoun had been grappling with the question of the agricultural potential of the Peace River country, Dominion land surveyors were reporting that the area of arable land along the fertile belt was much greater than had been previously estimated.[80] These findings probably served only to heighten the Professor's sense of being on the right track, and he now saw it as his role to set the record straight regarding the southern grasslands. The confidence and seriousness with which Macoun undertook this duty was demonstrated in May 1878 when Fleming sent him a copy of the commissioner's report of the westward march of the North-West Mounted Police march westward in 1874 along the forty-ninth parallel. 'I utterly dissent from Col. French in everything except his remarks on the country in the vicinity of the Boundary,' he replied. 'I think I see my way to annihilate French. I will touch him on his want of knowledge and also on his want of honesty.'[81]

This self-styled image as the saviour of the much-maligned prairie district was confirmed upon Macoun's arrival in Winnipeg in early June 1879. He was personally welcomed by Premier Norquay, made a member of the Manitoba Historical and Scientific Society and even approached by speculators. 'We start today at noon ... and from present appearances our expedition must be a success,' he wrote to his wife on 10 June 1879. 'I have no fears for the future. Dr. Schultze [sic] told me yesterday that if I did as much for the government this year as I did in 1875 I would be regarded as a public benefactor.'[82] He sent a similar letter from Portage la Prairie: 'It is more than likely my expedition will cause a sensation upon my return. Should I succeed as I expect my promotion is sure.'[83] As he prepared to head out across the Great Plain, then, it was as if the hand of destiny were on his shoulder. Certainly, he was blessed with favourable weather. Rainfall records for western Canada indicate that the 1870s were the wettest decade in the nineteenth century.[84] These conditions are corroborated in another of Macoun's letters home, this time from Fort Ellice on 19 June 1879: 'You spoke in your letter of rain. We have had it. 11 hours on Monday week. 36 hours on Friday & Saturday. A thunderstorm yesterday and today it is falling in torrents.'[85]

Macoun spent the next five months in the field. From Fort Ellice, his five-man party followed the Carlton trail as far west as the 102nd meridian where they left the trail and struck out by compass for Long or Last Mountain Lake. Here, on 12 July 1879, in recognition of the 'Glorious Twelfth', they decorated their horses with the reddish-orange lilies of the area and marched southwestward to the beat of an old tin pan. At the Elbow of the South Saskatchewan River, they checked the feasibility of building a canal to carry water from the Saskatchewan into the Qu'Appelle River – part of the former Mackenzie government's scheme to connect British Columbia with the rest of Canada by a series of rail and water links. They then travelled north to Battleford, the capital of the North-West Territories, to pick up the balance of their supplies before proceeding southwestward across the great treeless plain of the second prairie steppe. They reached Fort Calgary one month later. After a brief exploration of the Bow River Pass area, they headed north to Edmonton, where Macoun telegraphed a preliminary report to Ottawa[86] before starting eastward for Winnipeg, via Battleford.

On 2 October 1879, with Fleming in attendance, the Conservative cabinet reviewed the summer's field surveys and officially endorsed the Yellowhead route for the railway. Macoun, meanwhile, had not disappointed Tupper. His 1880 report of his activities was a highly favourable overview of the land, wood, and water resources of the great plain. 'I am quite safe in saying,' he predicted, 'that 80 per cent of the whole country is suited for the raising of grain and cattle and would not be the least surprised if future explorers had found a more favourable estimate. Only two points in the country explored were noted where it was probable the rainfall was too light for the successful raising of cereals.'[87] Macoun reached this conclusion about the region's potential on the basis of a number of diverse yet interrelated field observations. Crossing the great plain between Battleford and Calgary, he noted that the best tracts of land were devoid of timber, whereas the dry, sandy hills of the region were covered with wood. He also discovered thick swards of luxuriant grasses in areas where the buffalo herds had once been so enormous that Palliser had complained about the lack of feed for his horses. These rich grasses also grew in sandy or gravelly soil, but not where the heavy cretaceous till was the prevailing surface material. There, saline plants dominated. It was evidently too moist on other parts of the plain for them to exist. Finally, despite frequent thunderstorms, he found that streams or ponds were a rare occurrence in the heavily grassed areas. Rains that fell on the cretaceous till, on the other hand, could not penetrate the baked, crust-like surface and collected in alkaline sloughs or lakes.[88]

These findings led Macoun to conclude that the great plains environment was treeless because of the annual prairie fires that swept freely over the region, not because of limited precipitation.[89] The peculiar ecology of the cretaceous-till areas further suggested that the region's apparent aridity was a simple matter of surface cover. 'I, therefore, make the aridity where it exists one of soil and not climate,' he reasoned in his report. 'All the arid spots, all the salt lakes and the brackish marshes of the entire plain were traced to one cause – the Cretaceous clay.'[90]

As with his earlier work on the climate of the Peace River country, these ideas on the cause of the prairie's barrenness did not originate with Macoun. James Hector, the geologist on the Palliser expedition, had been initially puzzled by the sparse vegetation of the grasslands when 'there [was] quite a sufficient amount of moisture in the atmosphere during the summer months to support a more vigorous vegetation.'[91] He subsequently decided that the 'desert' country derived its character from the nature of its soil; the prevailing cretaceous clay baked under the hot summer sun into a hard impenetrable surface.[92] George Dawson greatly expanded on Hector's thoughts; his 1875 report contained the suggestion that the prairies originated from the frequent passage of fires rather than from rainfall deficiency.[93] He also placed emphasis on the parent material of the soil and its inherent fertility or aridity. He noted that glacial drift produced more fertile soil, particularly in those areas underlain by impervious rocks or clays that retain the water level near the surface. On the other hand, areas with a soil and subsoil too light to retain moisture were best suited for pasture land, while those based on cretaceous till were condemned to sterility.[94]

Macoun's 1879 conclusions were therefore not in any sense new.

What differed, however, was the way in which they were presented to the Canadian public. His experience with the Mackenzie administration had taught him the value of publicity; although he had unlocked the secrets of the Peace River district, he had largely been ignored by the government of the day. He was not, then, going to take any chances with the Macdonald Conservatives, even though his findings appeared to suit the needs of the new government. To guarantee his success, he began to cultivate an image as the country's leading expert on western Canada by embarking on a series of public talks over the next year and a half.

The first of these talks, entitled 'Our Wondrous West,' was delivered to an overflow crowd at the Winnipeg city hall on the evening of 20 November 1879. In what would become one of his characteristic disarming tactics, the professor cautioned his audience that he appeared before them 'with fear and trembling' because they might not believe his statements. He then went on to describe in an entertaining fashion his exploits during the past summer, leading up to the inevitable conclusion that Canada had inherited a land of 'illimitable possibilities.' He rarely mentioned the existence of bad land, and, when he did, suggested that 'in comparison with the enormous extent of that great region, you can't think of a few sand hills.' In fact, in keeping with his aim to advance his public reputation, he boldly announced that, 'a year hence I may be able to tell you of other and just as valuable tracts which I myself shall have explored.'[95] The lecture certainly had the desired effect, for it was printed verbatim the next day on the front page of the *Manitoba Free Press*.

That winter, while working in Sandford Fleming's Ottawa office, Macoun lectured to crowded halls and churches throughout southern Ontario. A variation of his Winnipeg speech, these talks were in most cases designed to entertain as much as to inform. When invited to address the prestigious Canadian Institute, however, he drew upon existing meteorological studies[96] to prepare a more formal lecture that combined his earlier work on temperature with his most recent investigations on rainfall. The tone remained strident though. At the outset of his speech, he chided his learned audience for continuing to cling to the 'old prejudice' about the southern country, implying that the region's true value would be evident if they only opened their eyes. 'We should be the means of enlightening the world as to the extent of the resources of the "Great North-West,"' he lectured, 'and in so doing, possibly of acting as special agents, fulfilling the beneficent intentions of the all-wise Creator.'[97] In what probably appeared as an ironic twist, the Professor then suggested that the large desert plateau within the United States actually exerted a beneficial influence on the climate of the western interior. The inland desert, according to Macoun, caused warm, moisture-laden air from the Gulf of Mexico to curve northwest-ward and disperse over the Canadian plains. This unique wind pattern resulted in an exceptional climate. Not only were the summers characterized by a high and uniform distribution of heat throughout the entire region, but much of the rainfall came during the growing season and ceased at harvest.[98]

Macoun's comments about the seasonal distribution of moisture were essentially correct, in that precipitation, although scanty, usually comes at the critical point in the growing season. They were mislead-ing, nonetheless, in that, from place to place or from year to year, rainfall in western Canada is subject to frequent wide deviations from average seasonal distribution. The south presents the greatest variabil-ity, and long, dry spells are not uncommon. To acquire experience in the plains region and be able to express a competent opinion consequently required detailed study over a number of years, not just one field season – particularly one that was abnormally wet.

Despite these shortcomings, Macoun's statements did not provoke a storm of criticism. The idea of a desert in the southern portion of the western interior was a disturbing notion that Canadians would quite happily have dismissed outright. It is quite easy to understand, then, why the Professor became such a prominent figure. A respected man of science, he not only seemed to have deciphered the physical geography of the North-West, but put his findings forward with such evangelical fervour that it was difficult not to be converted to his faith. In fact, the lone dissenting voice at this time was that of Charles Horetzky, the botanist's travelling companion in 1872. Still strongly committed to a northern rail route through the Pine River Pass, the temperamental Horetzky was greatly distressed by what he interpreted as Macoun's abandonment of the northland in favour of the prairie region. He consequently tried to discredit the Professor in a May 1880 pamphlet, *Some Startling Facts Relating to the Canadian Pacific Railway and the North-West Lands*. He argued that Macoun could have personally

seen only a fraction of the land that he lauded, referred to his acreage
estimates as an array of imaginary figures, and accused him of
unabashed plagiarism of other explorers' reports. 'It cannot ... be
doubted,' he concluded in reference to Macoun's future surveys, 'that
the Dominion will be further enriched by many more millions of acres
... that another scientific adjustment of the map will be in order, and
that much of the arid, cactus region north of the boundary line will be
forever obliterated.'[99] It was a telling criticism. Horetzky, however, had
either just resigned or been dismissed from his CPR position and his
pamphlet attacked not only Macoun but also his former boss, Fleming.
His so-called 'startling remarks' were therefore dismissed as the
diatribe of one who refused to accept the final route selection
gracefully.[100]

Tupper, on the other hand, entertained no such doubts about the
reliability of Macoun's field work. Before the results of the 1879 field
parties were released, he reported, on 3 March 1880, in the Commons:
'It will be found that, instead of having overrated the character of the
country, the most sanguine views in relation to the fertility of the Great
North-West will be more than borne out by the positive information we
will be able to lay before the House on the subject.'[101] A month later
when the railway report was tabled, he first underlined the need for
favourable assessments for the sake of the railway venture and then
confidently quoted from Macoun's latest acreage estimates to back up
his earlier contentions.[102] Privately, however, the reliability of Macoun's statements appears to have been secondary to the fact that they
nicely dovetailed with the Conservative railway policy. Although
Tupper did briefly question the Professor before making his April
speech, he also 'encouraged [him] to do [his] duty and stick to what [he]
conceived to be the truth.'[103]

Tupper was so determined to demonstrate the potential value of a
land grant in western Canada that Macoun was once again dispatched
to the southern prairie district for the 1880 field season. This time,
however, he was given a zig-zag itinerary that took in the areas south of
the Qu'Appelle and South Saskatchewan rivers – areas that had been
condemned in earlier reports. Macoun accepted the assignment with
alacrity, confident that it would prove 'eminently successful.'[104] Starting from Brandon in late June, the Professor and his four-man party
proceeded westward amid steady downpours over an almost perfectly

level short-grass prairie to Moose Mountain. But as they moved out across the Souris Plain, they came across a broken, clay plain with little surface water and animal life but surprisingly tall green grass. With the location simply given as 'somewhere in the "Great Lone Land,"' Macoun described this abrupt change in a 12 July letter to his wife: 'On Friday morning we entered on this plain and only saw water once at sundown after we had made 27 miles ... The surface of the ground was so rough that we thought our carts would have been shaken to pieces. Some of the cracks were 2 feet deep and the whole surface was constantly uneven.'[105] Continuing westward towards the Missouri Coteau or third prairie steppe of present-day southern Saskatchewan, water became so scarce that the party was forced to swing north to Moose Jaw Creek before heading over a sandy, at times gravelly, land to the Old Wives Lakes. Here, they sought relief from 'a week of fearful heat' and in 'water ... just like brine,' took 'a delightful bath ... and were greatly refreshed.'[106] They then worked their way over a rolling country that alternated between badlands and well-watered pasture to the eastern edge of the Cypress Hills.

Any sobering influence that this difficult march might have had on Macoun's enthusiasm was quickly nullified by his findings on 13 August at a farm located thirty miles northeast of the Cypress Hills on a branch of Maple Creek. An unassuming Indian had sown the seemingly worthless cretaceous till in late May and, despite a long June drought, his crop was astonishing. Wheat and potatoes were growing in the same field with cactus and sage-brush. Writing to Sandford Fleming later that day, Macoun excitedly related his latest finding:

The farmer – Mr. Setter had ploughed up Sage Brush and Cactus and had sown wheat, barley and oats and planted potatoes and here on this *Arid* spot were all there quite green – too green for this time of year – and the land which remained unbroken so dry that it was impervious to rain while the cultivated! – it had been ploughed once 2½ inches deep – land received the rain and admitted it into the soil. I took a spade and dug into the subsoil quite easily and found it moist but the unploughed land I could not penetrate without great difficulty the surface was so dry ...

I confess that this last discovery has again unsettled my views regarding this country and I am now prepared to take even *higher flights* than any I have taken before. The matter of soil and rainfall may now be left out of the question. Fuel is really the great question of the future and if the lignite deposits turn out well the limit to the extension of settlement and the production of grain southerly is one that no one now can predicate but of this I am sure that *I* am far within the truth.[107]

Despite this unsettling experience, Macoun was not prepared to go as far as to suggest that all land was suitable for cultivation. 'I do not wish you to think,' he continued, 'I am praising this region as I believe less in the future of the Cypress Hills and the country around them ... both on account of soil and climate but I am thinking of those so called deserts possessed of fine soil but today are devoid of shrub or tree for many miles and which at this season the grass is mostly dried up and which on this account is classed by me and others as irreclaimable deserts.'[108]

This cautious optimism did not pervade the report of his field activities that Macoun submitted to the government later that fall. Granted, he did admit that water was so scarce as his party crossed the Souris Plain that they had located only four water-holes in a fifty-five-mile stretch. And he did report that the Cypress Hills district was particularly poor. Yet on the whole the general tone of the report was exceedingly positive. In a thinly disguised attack on Palliser's and Hind's earlier work, he stated that 'the appearance of the country passed through was altogether different from what I expected, having been led to believe that much of it was little else than desert.'[109] In fact, after recounting Setter's successful cultivation of the seemingly barren 'cactus sods,' Macoun argued that 'all the land where not covered with sand or gravel would yet "blossom like the rose."'[110] He also challenged the long-standing assumption that settlement should be initially restricted to the wooded fertile belt. In his 1879 report, he had raised the possibility that 'future settlers will prefer the prairie as there is less broken land, less marsh and swamp and less labour required to make a home.'[111] His 1880 expedition, however, had effectively removed any remaining doubts about the value of the southern region and he now extolled the virtues of the treeless plain. 'Experience has taught me,' he argued in direct contradiction to traditional ideas about good agricultural land, 'that wherever trees and brush are growing to look for a broken country and one that contains *too* much water, while the open treeless prairie generally condemned to sterility is by far the best farming land.'[112]

These ideas were taken to the extreme in the public lectures that

Macoun continued to deliver during the winter months. He had always tended to be somewhat reckless in his public assertions, but now they bordered on the absurd. In one of his typical speeches at the Hamilton Collegiate Institute, he presented himself as the sole authority on the North-West and defied anyone in the audience to contradict anything that he had said or written. He then dismissed in an offhand, at times humorous, manner any doubts about the region's potential. He stated that he had never seen a bad crop in the North-West, regardless of the character of the soil. He also reported that the western climate was at its worst and would improve with settlement. His most amazing remark, however, was his declaration that 'there [is] no such thing as the fertile belt at all – it [is] all equally good land.'[113] This same lecture created such a stir in Ottawa that it was repeated with lantern slides and plant specimens at the special request of the governor-general.

Macoun's idea that western soils would become increasingly productive once the virgin prairie land had been broken and brought under continued cultivation appears to have been borrowed from the American frontier experience. As American farmers pushed out beyond the Mississippi Valley during an unusually wet period following the Civil War, there was a widespread belief that precipitation levels on the plains were actually rising with agricultural settlement.[114] Macoun, during his travels westward via the United States, had not only probably been made aware of this idea that ploughing increased the effectiveness of rainfall but also had seen at first hand how the American farmer was proving the plains to be fruitful. In fact, he often spoke of the 'practical Yankee' who 'can raise the greatest amount of wheat with the least possible expenditure of labor ... on the open prairie.'[115] Like Macoun's other statements on western Canada's physical geography, however, this 'rains follow the plough' notion was fraught with difficulties. By attributing equal capabilities to all the soils of western Canada, he created the false impression that bringing the shallow and light soils of the second and third prairie steppe into agricultural production was a simple matter of cultivation. In fact, they were less drought-resistant than the other western areas and required special techniques. Otherwise, serious agricultural problems resulted: carelessly broken prairie sod would erode and blow away, while successive crops would draw off moisture and eventually suffer from severely reduced yields.

These problems underlying Macoun's sweeping generalizations

would become evident once settlers began to farm the open prairie. What mattered right now, however, was that the Professor's recent findings perfectly suited the needs of the Canadian government. On 21 October 1880, the Macdonald Conservatives had finally signed a Pacific railway contract with the Canadian promoters of the highly successful St Paul, Minneapolis and Manitoba Railway. Under the terms of the contract, the syndicate agreed to build within ten years an all-Canadian transcontinental line for, among other things, twenty million acres of land 'fairly fit for settlement.' During the debate on the proposed contract, Charles Tupper drew particular attention to Macoun's 1880 exploration to allay opposition concerns that much of the best land in western Canada would fall into the syndicate's hands; only a small portion of the fertile lands would be absorbed by the railway land grant. 'Now we find that Professor Macoun,' the minister advised the House, 'found that that great Missouri section of barren country which was supposed to extend into Canada in the Northwest, was in great measure valuable and fertile land. He found that the idea that it was a desert was an entire delusion and that instead of that a great portion of these lands ... are largely fit for settlement, and they are included in the contract in the lands "fairly fit for settlement."'[116] The Professor's field work also seemed to have figured in the wording of the railway act that was passed into law on 17 February 1881. According to the legislation, the railway land grant was to be secured within the fertile belt, 'that is to say the land lying between parallels 49 and 57 degrees north latitude.'[117] The boundaries of this enlarged fertile area just happened to correspond with those in Macoun's 1879 summary of the lands available for settlement.[118]

Although Macoun's findings during the 1879 and 1880 field seasons, together with the signing of a railway contract, effectively made his work for the government complete, he was sent west again in the spring of 1881 to examine the eastern slopes of the Duck and Porcupine mountains, just west of the Manitoba lakes region. The assignment represented a complete change of pace from the past two summers' work, for the Professor was to assess a relatively small area and, as his travelling companions, he took three adolescents – his eldest son, James, George Moore of Winnipeg, and Henry Williams of Belleville. The survey was revealing, nonetheless, in that it showed the degree to which Macoun was now prepared to overlook or dismiss the

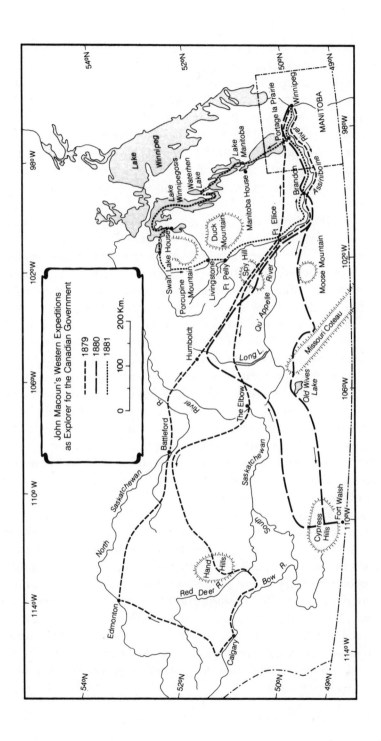

John Macoun's Western Expeditions
as Explorer for the Canadian Government

- - - - - 1879
— — — 1880
············· 1881

0   100   200 Km.

physical difficulties of a region in favour of extending fertility over the entire North-West. Conducting his exploration almost exclusively by water, supplemented by short excursions inland or observations from treetops, Macoun found the lands bordering Lake Manitoba and Lake Winnipegosis and along the Assiniboine River thoroughly saturated with water. Yet he could not bring himself to submit an unfavourable assessment. 'All the soil is good,' he reported in an attempt to downplay his discouraging findings, 'and all that is needed is a gradual clearing and drying of the land, and there will be no richer region in the interior.'[119]

This tendency to praise each region as it was explored by him was a consequence of Macoun's almost blind faith in western Canada and its capabilities. He was overwhelmed by the immense size and seemingly limitless resource wealth of the North-West during his first exploratory survey in 1872, and each successive trip served only to intensify this enthusiasm. He did not see the western interior as a great natural history laboratory in which to test the validity of Darwin's theory but rather as an untried, somewhat sacred agricultural paradise where man would experience a fresh start. 'A look over a field of growing grain,' he later wrote during one of his more reflective moments, 'is all that is necessary to tell the practical man that here is a land with untold wealth in its soil, and as the life giving breeze fans his cheek, he feels that here life means an unending pleasure. The blood courses through his veins as it did when he was a boy, and he is young again in spirit if not in years. The sensation is irresistible, and all men feel never to be forgotten pleasures as they gaze on the waving fields of grain and prairie grass.'[120]

Macoun's evaluation of western Canada's potential was also influenced by personal ambition. Not content with the prospect of being little more than an immigrant farmer, he turned to botany for self-advancement as much as for self-fulfilment. He dearly wanted to be appointed to the Geological Survey so that he could pursue his botanical studies while attaining the status and recognition that he craved. He consequently tried to be as accommodating as possible in his reports, knowing full well that his chances of advancement depended upon positive, practical findings. He consoled himself, in the interim, with the public acclaim that he garnered by discrediting earlier pessimistic assessments.

Finally, the Professor's sweeping statements were based on the seemingly successful application of his skills as a naturalist. Extremely confident in his abilities, he genuinely believed that where other investigators had failed, his field work had scientifically demonstrated that the Peace River country and the southern plains were equal to the fertile belt. These discoveries, in turn, instilled in him a dangerous sense of infallibility. He did not seem to realize that the application of his sectional and seasonal observations to the vast western interior led to misconceptions and distortions. His so-called revelation of the secrets of the North-West was not necessarily wrong – his characterization of Palliser's and Hind's work as 'a hasty conclusion'[121] was quite valid – but his own findings represented only the ideal aspects of the region's character. By concentrating on observation, and by generalizing from it, Macoun, like the investigators before him, achieved a limited kind of truth.

Macoun's enthusiasm had its costs. He was guilty of encouraging settlers beyond the danger zone of aridity, and giving them a false sense of security. His statements about the great potential of the region suggested that the standard 160-acre homestead allotment was appropriate in grassland areas and that the settler required little assistance at the pioneering stage, especially when 'all that is needed is a mere scratching of the soil.'[122] The consequences were anxiety, disillusionment, and ultimately failure. In fact, one historian has suggested that federal homestead policies were a 'colossal national blunder' in that the western farmer 'had been permitted and encouraged to place himself in an impossible situation.'[123]

Macoun's championing of the agricultural worth of the open plains was not responsible, however, for the last-minute abandonment of the Yellowhead route for the Pacific railway. The decision in spring 1881 to reroute the main line through the southern grasslands was based on the syndicate's determination to secure all the potential traffic of the North-West by eliminating possible competition from American-based railways. Since the main line could serve or control only a limited area, the CPR builders decided to crowd the international boundary as closely as would be permitted by the federal government. This attempt to exclude American inroads into western Canada was further necessitated because the railway was to be an all-Canadian route running through the wilderness north of Lake Superior. Western traffic had to be secured to support this non-revenue-producing region. The syndicate therefore thought primarily in terms of national strategy, not local conditions; it would have built across the prairies even if the area had been poorly regarded, relying on branch lines north into the fertile belt. Macoun's role, in the end, was largely limited to providing the agricultural justification for a route that was selected for essentially other reasons.[124]

The Professor performed a somewhat similar role for the federal government. Canada had acquired the North-West in 1870 on the simple assumption that much of the region would soon become the future home of countless millions. Macoun confirmed this expansionist vision in dramatic fashion; he not only claimed that temperature, rainfall, and soil were suitable for large-scale agricultural development, but couched his findings in terms and images that matched Canadian expectations for the region. This application of his naturalist skills to the great nation-building tasks at hand made him the darling of the Conservatives. It was for this reason that Macoun finally in November 1881 secured the elusive position of botanist to the Geological Survey of Canada. It is extremely doubtful whether such a position would have been created had the Professor not performed such a valuable service for the Canadian government in western Canada. Ironically, the man who confirmed the appointment was the same man who, during the debate on the 1877 Survey Act, had questioned the idea of undertaking what he regarded as 'merely subsidiary and incidental inquiries.'[125] It was now up to Macoun to show how wrong Macdonald had been.

Plants and Politics

John Macoun's appointment as Dominion botanist coincided with the beginnings of a troubled period for the Geological and Natural History Survey of Canada. The national development policies of the Macdonald Conservatives were based on the widely held assumption that Canada's natural resource wealth was unlimited and that, if these resources were made 'useful,' the country's future as a great nation was assured.[1] Ottawa consequently expected federal scientific agencies to provide simple, straightforward practical information that could be immediately used by private interests to unlock the resource wealth of the young Dominion. It did not believe in science for its own sake but rather in the use of science to further economic growth and promote national strength and prestige – what one intellectual historian has described as 'an entrepreneurial scientific ideology.'[2]

This idea of science lay behind the Conservative decision to implement one of the provisions of the 1877 Survey Act and transfer the Geological Survey from Montreal to Ottawa in the spring of 1881. It also led to a certain degree of friction between the Survey and the federal government. Throughout the 1880s, parliamentarians of both political stripes were forever criticizing the federal agency for not doing enough to promote the material well-being of the country and felt no misgivings about interfering in its organization and operations. They generally did not accept the argument that scientists should be left on their own to run the Survey or that appointments should be made on the basis of knowledge and expertise not political persuasion. Since the government provided the funds for the Survey, they insisted that they had every right to demand certain returns from their investment and they expected positive, constructive information.[3]

Strangely enough though, during his first five years as Dominion botanist, Macoun was largely engaged in the study of the native flora of Canada rather than its practical applications. This focus is somewhat surprising, given the Conservative government's overriding concern with the exploitation and development of Canada's natural resources. One would have expected the Professor to carry out surveys in other parts of the country similar to those that he had performed in the North-West. Macoun's direction is even more remarkable in light of the widespread feeling that the Survey under Selwyn's leadership was concentrating too much on pure science and was getting away from its original function of examining and aiding in the development of the mineral resources of the country. Yet even though Macoun became personally embroiled in the debate over the Survey's purpose, his own botanical work was never questioned. Because of his reputation as the discoverer of western Canada's true potential, he was above reproach. It was simply assumed that his cataloguing of Canada's plant life under the Survey's auspices was a continuation of the pioneering work that he had conducted in western Canada during the 1870s. Besides, Macoun continually talked about the practical applications of his botanical knowledge and, whenever he was asked to comment on the potential of a region, he did so in such an overconfident manner that he created the distinct impression that here was a scientist whose work was not to be questioned.

John Macoun had no difficulty measuring up to Ottawa's expectations of a scientist in its employ. His practice of simply listing and describing the flora of a particular region, as opposed to the testing of Darwin's work, was precisely the kind of science that could be understood and therefore appreciated by federal politicians.[4] His findings during his five western exploratory surveys, moreover, perfectly mirrored the philosophy underlying the government's development policies. He was essentially of the same mind as his new political masters. Almost immediately after securing his appointment to the Survey, Macoun brought together his work over the past ten years in the extremely popular *Manitoba and the Great North-West*. Whereas his past reports could have been used to justify any route for the transcontinental railway, he now tried to help the CPR cause by popularizing the southern plains at the expense of the previously heralded fertile belt. He also adopted a highly critical tone towards Palliser's assessment, while portraying the potential of the western interior in the most exaggerated terms to date. 'It is extremely difficult for one unacquainted with the subject to grasp the extent and capacity of the country,' he wrote. 'It is practically boundless as far as this generation is concerned, and long after our bones are mouldered into dust there will still be millions of acres untrodden by the foot of the husbandman. Pages could be written at this point filled with what would seem highly colored and extravagant adjectives, and yet these would not reach the reality.'[5]

Finally, Macoun had no qualms about government interference or about helping the Canadian business community. A staunch Conservative who had learned the value of political influence in the 1870s, he wanted to serve the interests of the government, expecting rewards in return. He also was intent on maintaining the public profile that he had generated during his western work and was prepared to bring his excessive optimism to the aid of Canadian entrepreneurs and developers, as well as the imperial connection. 'I know the country better than any other living man,' he had written to the deputy minister of the interior the year before his appointment, 'and I feel it my duty, in the interests of Britain, to say to those whose patriotism is not a mere sentiment, that England's future is more intimately connected with Canada than to-day appears on the surface.'[6]

In the spring of 1882, however, Macoun had more immediate concerns. Waiting in Belleville, he had not received any official notice of his new position. Nor had he been paid a salary since the end of December. To clear up the matter, the Professor tried to secure a copy of the order-in-council from Lindsey Russell, the deputy minister of the interior who had arranged his appointment to the Survey; but he returned home empty-handed. He then turned to his old friend Mackenzie Bowell. Noting how he had 'done more to bring about the present state of affairs in the West' and how he 'thought he had surmounted all difficulties,' he asked the minister of customs whether he could help end 'the state of uncertainty.'[7] Yet when Bowell spoke to Macdonald about the matter, the Prime Minister confessed that he had forgotten the circumstances surrounding Macoun's appointment and that he would be guided by his deputy's actions.[8] Russell, in the meantime, had been in touch with Dr Selwyn and had straightened things out. Effective 1 January 1882, Macoun was botanist on the Survey temporary staff at a salary of $125 per month. As for the order-in-council making the appointment official, it never came before council and the Professor was paid out of the Survey's contingency fund until he was made a permanent officer with a first-class clerkship on 1 July 1883.[9]

Macoun's appointment probably came as a pleasant surprise to the Survey director. After his unsuccessful efforts to try to secure a position for the botanist in 1874–5, he had more or less given up on the idea. 'If promises are fulfilled,' an exasperated Selwyn had advised Macoun in late December 1880, 'there would be no difficulty about your appointment on the Survey.'[10] When he consequently learned from Lindsey Russell that the appointment had finally been made, he lost no time in contacting the Professor and instructing him to begin preparation of a catalogue of all the known plants of the Dominion with particular emphasis on their distribution. It was hoped that the next logical step, a flora of Canada, would follow.

That Selwyn would give Macoun such an assignment at a time when politicians were demanding practical resource data clearly demonstrated his determination to make the Geological Survey a respected organization in the world scientific community. The project also underscored the extent to which any distinctively Canadian enterprise in the natural sciences had been slow to develop. Although the first work on Canadian botany, J.P. Cornut's *Canadensium Plantarum Historia*, had been published as early as 1635, botanical activity in seventeenth- and eighteenth-century Canada was a foreign-directed enterprise, chiefly concerned with the collection and description of the flora of the New World and their comparison with European species. Any unique local study was carried out by amateurs such as clergymen and physicians who made simple, crude collections.[11] This dependence upon foreign investigations for a knowledge of Canadian plant life continued well into the middle of the nineteenth century. In 1840, the eminent British botanist Sir William Hooker drew largely upon the collection efforts of naturalists attached to British Arctic expeditions to produce his two-volume *Flora Boreali Americani*. Hailed as the most complete summary of the flora of British North America to date, it was nonetheless a byproduct of European activity. Few Canadians, however, were to be found working on their flora at this time. There was little organized knowledge of Canadian plant species on even a regional – let alone national – level; nor was there much correspondence or exchange of specimens among Canadian collectors. As Dr George Lawson, the Scottish-born professor of chemistry and natural history at Queen's University, observed, 'Botany is at a low ebb in Canada, at a lower ebb than in most civilized or half-civilized countries on the face of the earth.'[12] Biological sciences, in fact, were so little developed that it was possible to master several fields at once, to be a professor of 'Things in General.'[13]

This inactivity came to an end in the 1860s when Canadians finally moved to bring the study of botany under their control. A time of national awakening, fuelled by the movement towards Confederation, there were no less than half a dozen attempts under way in different parts of British North America during this period to prepare a flora of Canada. Only one of these projects, however, came to fruition – Abbé Léon Provancher's two-volume *Flore Canadienne* (1862) – and it was largely ignored because it was considered premature. Unfortunately, the botanist who was probably most capable of producing a respectable flora, the Abbé Louis-Ovide Brunet of Laval University, was apparently discouraged from bringing his own manuscript to completion by Provancher's efforts.[14]

Macoun was undoubtedly aware of these various efforts to produce a flora of Canada and, in a few instances, was called upon for specimens and lists. As his chances of securing a government position dimmed in the mid-1870s, however, he also took up the challenge. Confident that his wide-ranging collecting efforts, particularly his 1872 and 1875 western expeditions, placed him in an ideal position, his first step was the publication of a checklist of the various plant groups of the Dominion.[15] With this in mind, Macoun made the trip that most botanists of his generation did at one time or another in their careers and visited America's foremost botanist, Dr Asa Gray, at Harvard University in September 1876. Gray's retirement project, *The Synoptical Flora of North America*, effectively required him to try to see all the collections made in North America. The Harvard herbarium consequently functioned in the 1870s 'as the clearing house of all botanical information.'[16] Macoun had been tied into this network since his return from the Fleming expedition, regularly sending to Gray or Gray's assistant, Sereno Watson, instalments of Canadian flora for determination. In fact, Gray was prepared to fund a collecting trip to Lake Superior in 1875 if Macoun's position on the Selwyn expedition fell through. This informal instruction from one of the giants of North American botany was at times testy, for Gray was not inclined to accept Macoun's overconfidence. When, for example, in October 1878, the Professor forgot to include a particularly important specimen in a collection sent to Harvard, Gray tersely advised, 'Who is going to *believe* in it without seeing. *Not* yours truly.'[17] Despite such exchanges, Macoun's visit with Gray was one of the highlights of his botanical career. He felt privileged to be allowed to work in the great man's herbarium and came away impressed with the importance that Gray attached to exact identification.

Even after Macoun had secured the position as government explorer, he continued his preparatory work towards a flora of Canada. He made large plant collections during his three successive western surveys and, in exchange for a complete set of duplicates, repeatedly called upon Dr Gray to identify his more difficult specimens. He also updated and corrected his 1878 checklist and was close to issuing a revised version when he was appointed to the Survey.[18] Dr Selwyn's decision to have Macoun prepare a catalogue of Canadian plants was therefore a timely assignment – something that the botanist could pull together with little additional work. Yet instead of simply sitting down and producing a manuscript based upon existing published records and his own voluminous collections, the Professor immediately made plans for the 1882 field season. Two reasons lay behind this action.

During the course of his botanical studies over the past twenty-five years, Macoun had been frustrated on several occasions to discover that many of his specimens of a supposed species did not match the published descriptions. European names and descriptions had automatically been applied to North American species when they were actually different forms altogether. The same problem existed between American and Canadian species. The Professor therefore interpreted Selwyn's instructions as a challenge to try to clear up this 'mixing and muddeling [sic]';[19] he was determined to try to collect, or at least personally examine, every plant species peculiar to Canada. As he told Sereno Watson, 'I hope to get our botany in proper shape in the course of a few years and as I work the ground over I propose doing it so thoroughly that all additional work will be added to the general stock without having to go over the same general ground again.'[20] Macoun's investigations, moreover, had always been undertaken in the field and he was not about to change his methods. In fact, he was quite anxious to exploit the opportunities of his new position to search for species new to Canadian botany. 'The more I know of our flora, the more I am satisfied that we have numerous forms not found south of Lat 45,' he advised Dr Gray. 'When I reach ... the lower forms I will astonish you.'[21]

Macoun's first few years at the Survey were thus extremely active ones in the field. In the spring of 1882, after the matter of his appointment had been resolved, he collected for several weeks in southwestern Ontario. He then headed for the Gaspé peninsula and spent the rest of the field season working his way along the south shore of the St Lawrence River with a geological party under R.W. Ells. The following summer, he botanized in Nova Scotia and then on Anticosti Island, where the black flies were so bad that he spent the entire visit encased in a local mixture of tar and castor oil. Despite such irritations, Macoun never wavered from the task he had set himself, and by each season's end he had amassed the large collections that were the trademark of his field activities. Such activity also increased his confidence in his own collecting ability. 'Circumscribed as you are in England, and with hosts of collections in all branches of the subject,' he informed Sir Joseph Hooker of Kew Gardens, 'you cannot have the faintest idea of the vast amount of bodily toil and close study necessary to become familiar with nearly 4,000 species and collect nearly all of them in their natural haunts.'[22] Canadians received similar lectures on his expertise. Giving evidence before the Commons Committee on Agriculture and Colonization after his first season in the field as Survey botanist, he boldly announced, 'I profess to know every species of plant that has been found in the Dominion of Canada, from the Atlantic to the Pacific to the extreme north, and what is more, I know the range of every species.'[23]

Macoun was equally busy following his move from Belleville to Ottawa with his wife and five children in November 1882. Because of the decisive role he had played in the debate over the fertility of western lands, the bearded professor with the Irish accent and infectious enthusiasm was no stranger to the city, and he quickly became a prominent figure in social and intellectual circles. Indeed, he arrived at a time when Ottawa was trying to shake off its image as a raw lumbering town and become a city more in keeping with its status as the Dominion's capital.[24] This mood matched that of Macoun; he had been anxiously awaiting his appointment for several years and looked upon his relocation to Ottawa as the long-sought opportunity for bigger and better things. He consequently bought a fairly substantial brick house on James Street, thirteen blocks south of the Parliament buildings in the respectable Wellington district and within walking distance of work. He also became one of the leaders of the Ottawa Field-Naturalists' Club, as well as attending the meetings of the Literary and Scientific Society whenever possible. Through his association with some of the country's leading scientists, in particular Principal Dawson of McGill, he was also named a charter member of the Royal Society of Canada —a body established by the governor-general, the Marquis of Lorne, as a kind of 'national showcase of Canadian talent.'[25] This honour was accompanied by the placing of Macoun's name on the visitors' list at Rideau Hall, and he and his family regularly attended the governor-general's skating and tobogganing parties.

When not in the field, however, the majority of Macoun's time was taken up with office duties. In addition to working up his own large collections, he received a steady flow of mail from amateurs across the country who were anxious to take advantage of his expertise. The soon-to-be-famous animal story writer Ernest Thompson Seton of Carberry, Manitoba, for example, reported, 'I find an endless fund of enjoyment and interest in the flowers about me, and my pleasure is only marred by my ignorance.'[26] Macoun gladly examined these collections, since they added to his knowledge of the distribution of species and could be cited in his Canadian plant catalogue. It was time-consuming work, nonetheless, for the material often consisted of 'scraps' of specimens and was 'so mixed up that *nature* herself could not unravel the tangle without *growing* the stuff over again.'[27]

Almost from the time of his appointment, Macoun was assisted in his Survey duties by his eldest son, James. Born in Belleville in 1862 and educated at Albert College, Jim or Jimmy, as he was commonly known, spent the better part of his life helping to further the Professor's career. He took great pride in his father's accomplishments and growing reputation in the 1870s and came to be relied upon during the elder Macoun's prolonged absences in the field or at Ottawa. Under his father's guidance he also became an intrepid naturalist in his own right. He was encouraged from an early age to work in the herbarium and, as his father had done during his own boyhood in Ireland, had his own flower garden, grown from seeds that had been specifically collected for him in the North-West.

Following the family's move to Ottawa, Jim joined the Survey's temporary staff in 1883 as a field assistant. Attached to various Survey parties during the summer months, he quickly proved himself an extremely valuable addition to the natural history section, often working under difficult and hazardous conditions in remote areas and bringing back whatever specimens it was possible to carry. When Survey palaeontologist T.C. Weston returned from a fossil-collecting trip to western Canada in 1884, he told Selwyn, 'Give me James every time. He knows no fatigue and fears no danger.'[28] During the winter, as he had been doing throughout much of his adolescence, the younger Macoun worked at his father's side, mounting, labelling, and arranging specimens. Yet apart from sharing a keen fascination with the natural world, father and son were different in many fundamental ways. The Professor was a self-righteous, outspoken, abstinent Conservative and an ardent imperialist, while the genial, unassuming Jim was a committed labourite and civil service reformer who liked his beer. Macoun never interfered, however, with his son's political beliefs; in fact, Jim often used the back pages of the office letterbooks to order socialist literature. What mattered most to the Professor were his son's skills as a naturalist. As for Jim, his own modest nature and absolute sense of devotion to his father and his work ensured that he would remain in the old man's shadow for the next thirty years.

Even though Jim handled the more mundane office work, thereby freeing his father's time, the Professor continued to send the better part of his field collections to foreign specialists for identification and verification. This practice was not unusual. Many Canadian naturalists relied upon European and especially American scientists for assistance and direction.[29] Nor, in Macoun's case, was it surprising. During his nearly three decades of botanical study, his expertise was as a collector – a phenomenal one, without question, but still a collector. As he admitted to Dr Gray, it was only 'with such an expense of time and trouble' that he worked out the names of most of his difficult specimens; even then, he was not sure that his determinations were correct.[30] It made more sense to Macoun, then, to send his questionable or potentially new specimens to recognized authorities for examination. In this way, he would not get bogged down in the winter months trying to work over his large collections and, in the process, possibly jeopardize the sharp eyesight needed for field work. 'I am satisfied to collect and discriminate generally,' he confided to American moss specialist G.N. Best of New Jersey, 'but I will not destroy my eyes by working with the microscope.'[31] He could also be assured that his collections were being attended to by some of the acknowledged experts in each particular order with the ability not only to identify his specimens but to describe species new to science; and if he was not satisfied, he could send them back for re-examination or to some other authority. The professor firmly believed that 'by this method ... I am on the right track to get at the truth.'[32] But it had a major drawback. Despite the presence of a full-time botanist on the government payroll, Canadian flora continued to be worked up by foreigners, in particular by Asa Gray and his associates.

Once correctly identified, Macoun's specimens, along with those collected by geological field parties, were added to the Survey herbarium, which had been created in 1883 with the purchase of Macoun's private collection. Initially, Dr Selwyn had expressed little interest in acquiring the Professor's herbarium, given the Survey's limited budget and the severe space restrictions at the Sussex Street headquarters. Macoun, however, was not simply satisfied to be on the Survey staff, occupying the position he had pursued for several years. He was convinced that his botanical work was comparable to that of the geologists and, just as he had promoted the North-West's potential, he was prepared to take matters into his own hands and do all he could to secure greater support and recognition of his studies. When a letter consequently appeared in the Ottawa *Citizen* in April 1883 criticizing the Survey's natural history collection, Macoun responded with one of his own in which he explained to the paper's readers that his herbarium was not wanted and that he was toying with the idea of selling it to a public institution in the United States.[33] Survey geologist and assistant director Robert Bell sided with Macoun three days later. Noting that the botanist's collection 'would be almost impossible to replace ... in this generation at all events,' he pointed out that, 'We have a Geological and

Natural History Survey and it is very desirable on account of its scientific and economic aspects, that the natural history part of the Survey should receive due prominence.'[34]

Such criticism was warranted. Although Survey officers regularly returned from the field with a wide assortment of specimens, natural history remained secondary to geological pursuits. The Macoun and Bell letters, at the same time, were an expression of the intense professional rivalries that seethed within Survey ranks. In Robert Bell's case, his comments were little more than a veiled attack on Selwyn's leadership. A member of the Survey staff since 1857, the irascible Bell believed that his Canadian birth and wide field experience made him a better qualified person than Selwyn for the directorship, and he was continually plotting to undermine Selwyn's credibility.[35] Macoun's target, on the other hand, was the assistant director, Joseph Whiteaves. The British-born Whiteaves had joined the Survey staff in 1876 as palaeontologist, but because of his experience as curator of the museum of the Montreal Natural History Society, he was also placed in charge of the Survey's collections. He continued in this capacity as museum director following the transfer of the Survey in 1881, and two years later assumed the position of zoologist as well. Macoun found this situation galling. He resented being under Whiteaves' supervision and frequently charged that the palaeontologist neglected his duties in the natural history area.[36] Yet, given the limited resources with which the Survey's natural history mandate was to be fulfilled, it is difficult to know what else Whiteaves could have done.

Lindsey Russell, deputy minister of the interior and the man largely responsible for the botanist's appointment, was extremely irritated by Macoun's and Bell's 'airing their views' about the sorry state of the Survey's natural history work. It was 'to say the least,' he wrote to Macdonald the day Bell's letter appeared,

very bad 'official form' and I think they merit a reminder that such is the case ... The first man [Macoun] is a good specialist and honest fool outside of that. The latter [Bell] is I think but a superficial scientist though a clever knave. It would have a wholesome effect on the consequence of the one and the intrigue of the other, if they were pulled up a little. Had my friend the Director a little more 'push' he would put a stop to unseemly exhibition ... Indeed, was it not for consideration of him ... I should as your deputy have done the needful by putting the extinguisher on their literary efforts ... though only trifles they are discreditable ones needing authority to prevent their recurrence.[37]

The Conservative government, however, was unable to contain the issue, which surfaced in the House of Commons the following month when the Survey's annual appropriation was being considered. Calling upon the government to purchase Macoun's herbarium, George Casey, the Liberal member for Elgin West, pointed out to Macdonald 'that sufficient attention has scarcely been bestowed to one of the purposes for which this survey was originally instituted ... of course it is exceedingly important to know what is underground, but it is of more pressing importance, as regards immigration to know what is grown naturally in every part of our territory ... I think at least half the money and energy of the Department should be devoted to explorations respecting botany and natural history.' This argument was supported by the new opposition leader, Edward Blake, who believed 'with great strength of conviction' that 'Canada ought to go very strongly into a natural history survey.' The prime minister, in response, neatly sidestepped the question by citing the lack of funds for such work, as well as reminding the House that the Survey had originally been established 'to define the geological position of the country.'[38]

There was little doubt from Macdonald's remarks that the natural history branch of the Survey was of low priority to his government. In fact, he was probably frustrated by the favourable support Macoun's work had received. The Conservatives were never happy with the 1877 Survey Act and believed that the agency required reforms of a more fundamental nature; in particular they expected the Survey to help realize the early development of Canada's seemingly unlimited resource wealth. Within a few months of their return to office, then, they began soliciting suggestions as to how to make the department more utilitarian. Thomas Macfarlane, mining engineer and discoverer of the famous Silver Islet mine on Lake Superior, was the most rigorous Survey critic, sarcastically noting in 28 January 1879 to one of Macdonald's colleagues that 'were it to be at once abolished neither government nor people would feel its loss.' He then went on to urge a complete reorganization in order to make the Survey 'a powerful instrument for stimulating sound *mining* enterprise and thus contributing its share to the success of a national policy.'[39]

This concern for efficiency and practicality had been the reason behind the transfer of the Survey and its museum to Ottawa in the spring of 1881. David Mills had defended the Survey bill during second reading in February 1877 arguing that, the department would never produce satisfactory results until it was placed directly under federal control and supervision.[40] Yet now, with the Survey right under Parliament's nose, its failure to provide positive, technical information was only magnified. Many politicians probably agreed with Macfarlane's private assessment that the 1877 act provided for little more than the Survey's removal to Ottawa.[41]

The Macdonald administration was thus determined in the early 1880s to make the Survey a development-oriented federal agency. In April 1882, in a move to bolster the Survey's activities in western Canada, the annual appropriation was increased by 20 per cent, to $60,000.[42] The following year, the government took the first major step in the agency's reorganization and transferred the salaries of the permanent officers to the civil service list, thereby making more of the annual federal grant available for field and laboratory operations. Because of this increased funding, politicians looked forward to exercising greater influence and control over the Survey. The improved financial situation appeared to have the opposite effect, however, when Dr. Selwyn began to hire a number of better-educated and more highly specialized scientists. For parliamentarians who viewed scientists as little more than collectors of useful facts, these new appointments seemed to widen, not close, the gulf between Survey objectives and government expectations. They also suggested that there was much truth to the general feeling that the Survey was more concerned with pure science than with resource exploitation.[43]

Interestingly, another reform the Conservatives considered but never actively pursued during this period was the dropping of the Survey's natural history duties. Scarcely six months before Macoun was appointed Dominion botanist, the prime minister was contemplating a draft revision of the 1877 act that struck the words, 'Natural History,' from the Survey title.[44] Macdonald perhaps reasoned that the department had enough difficulty providing practical geological information without worrying about Canada's flora and fauna, and that it would do better to concentrate its efforts in the area for which it had originally been established and that promised an immediate economic return. To his probable chagrin, however, the natural history branch, as evidenced by the favourable response to the Macoun and Bell letters in Parliament in April 1883, was one of the few Survey departments that enjoyed political approval and encouragement. This appreciation of the Survey's natural history work was undoubtedly the result of the Professor's widely-publicized findings in western Canada in the 1870s. It was also probably a consequence of the kind of science that Macoun practised. His methodical collecting and cataloguing of Canada's plant life, as opposed to wrestling with abstract theories, was not only seen to have practical applications but was also easily understood by the politicians who were called upon to endorse the annual Survey appropriation.[45]

Macoun used this support for his botanical work to resolve the question of what would be done with his private collection. Following the brief discussion of the issue on the floor of the Commons, he quickly prepared a petition outlining the absolute necessity of a good herbarium at the Survey. When he had secured the signatures of a number of senators and MPs, he then personally met with Macdonald on 23 May 1883 and received approval to negotiate with Dr Selwyn.[46] The Professor's herbarium at this time probably numbered around 30,000 sheets, but Selwyn wanted no duplicates. It was consequently agreed that the government would purchase only one sheet of every Canadian species from every province or territory in which it was found, as well as any form that differed slightly from the 'type' specimens that had been preserved. This stipulation meant that roughly one quarter of Macoun's herbarium was acquired at a cost of just under twelve cents per sheet. The remaining specimens, although kept at the office, were regarded as part of the Professor's private collection and were integrated into the main Survey holdings only after Selwyn had retired.[47]

While the Macdonald government tinkered with the Survey organization and responsibilities, problems surfaced within the department itself. Macoun and Bell were not the only letter writers at this time. On 6 April 1883, the Toronto *Mail* published a letter by Wallace Broad, a former Survey employee who had had a salary dispute with Dr Selwyn. Under the pseudonym 'Geologist,' Broad blamed the Survey's problems on the director, rather than on the lack of adequate funds.[48] Robert Bell, calling himself 'Bystander,' added his voice five days later in a letter that argued that 'the true "inwardness" of the institution ... should not be allowed to continue if the country is to receive anything like the return which it has a right to expect for the large sums annually spent on this Survey ... The fact is, the present director is utterly incompetent.'[49] Then on 13 April, the Montreal *Witness* entered the fray when it carried an editorial referring to the Survey as 'a dissatisfied department.'[50]

Selwyn, for his part, was certainly not well liked by his staff. He was an exacting, austere, though sensitive, administrator who expected unselfish devotion from his officers but dispensed few accolades in return.[51] He had, however, the support of the government – Macdonald had personally consulted him about changes to the Survey[52] – and, if he had defended his position and his policies publicly, he might have been able to disarm his critics. But his conservative nature and strong sense of propriety prevented him from doing so.[53] He simply expressed his concerns privately to Senator Sir David Macpherson, the acting minister of the interior. Macpherson was convinced that Bell 'has more to do with it any any other' and naïvely suggested that the two men settle their differences among themselves.[54] The grumblings from within the Survey continued, however, and on 24 February 1884 the Commons unanimously passed a resolution by Tory backbencher Robert Hall to appoint a select committee to investigate the organization.

The Hall Commission was not the first inquiry into the activities of the Survey. In 1855, the agency had received a blanket endorsement from a select committee of the Legislative Assembly of Canada appointed to review its record over its first dozen years of operation.[55] Nor were such inquiries unique to Canada. The United States Geological Survey was coming under criticism similar to that being levelled at its Canadian counterpart. Although initially established in 1879 to facilitate the exploitation of American mineral resources, the U.S. Survey, under its second director, J.W. Powell, had dramatically expanded its activities in the early 1880s to include other areas of scientific inquiry. This almost unilateral reorganization of the U.S. Survey away from an exclusive concern with economic geology greatly disturbed Congress and led to the appointment of the Allison Commission in 1884.[56] The American and Canadian surveys were therefore under investigation at the same time and ostensibly for the same reasons.

According to the Commons resolution establishing the inquiry, the Liberal-dominated select committee was to examine the current operation and methods of the Survey in an effort to determine whether more attention should be given to mining and metallurgical activities in Canada. Most of the questioning consequently tended to be of a critical, negative nature. Many committee members, moreover, could not resist the temptation to probe the internal situation at the Survey, and they found that several witnesses were quite willing to air their grievances. Such testimony served only to exacerbate the dissension within Survey ranks and, in the long run, undermined the effectiveness of the committee itself.

The Survey's natural history responsibilities were given only passing attention by the inquiry. Thomas Macfarlane, in his written submission, called for a significant increase in staff, including the addition of seven botanists,[57] while Robert Bell suggested, 'If the Geological Survey is also to be a Natural History Survey, it might be as well to have a qualified ethnologist and ichthyologist.'[58] In an apparent attempt to demonstrate the uneasiness that prevailed under Selwyn's leadership, however, Dr Bell tried to stir up tensions between Macoun and Whiteaves with his casual remark that the botanist was also understood to be naturalist upon joining the Survey but that the position had been taken away from him.[59]

Perhaps fearing that natural history might be sacrificed in the interests of more geological work by the Survey, Macoun was determined to avoid any such controversy during his brief appearance before the committee on 3 April 1884, and he played down any ill feelings that he held towards Whiteaves. He represented his relations with the palaeontologist as exemplary. He also supported Selwyn's leadership, testifying that the director had 'the interests of the country, and the interests of the Survey more at heart than any other member of the staff'[60] and that the current unhappiness stemmed from low salaries. As for Macoun's work for the Survey, its value seemed to be understood. Just one year earlier, when he had appeared before the Commons Committee on Agriculture and Colonization, two of whose members were now part of the Hall Commission, he had given a glowing coast-to-coast assessment of Canada's agricultural capabilities. There was consequently little questioning of Macoun's duties during his examination except from Conservative Simon Dawson of Algoma; and Dawson simply wanted to know whether the botanist would soon be examining the land that fell within the MP's riding boundaries. 'I am very glad you mentioned that,' the Professor replied, 'because I think it is important that a man like myself should be in that country. I could tell the climatic from the flora.'[61] In fact, the only controversial part of his testimony was his verbal sparring with one of the more assiduous committee members, Liberal Edward Holton. Whenever Macoun appeared before such hearings, he tended to be rather difficult and at times obtuse in his responses. Eventually, a frustrated Holton was forced to remind him, 'I am not here to answer questions, but to ask them and I will ask them too.'[62]

Macoun was the last witness during the month-long committee hearings. But because of the line of questioning pursued and the acrimonious nature of the proceedings, discussion of the situation at Survey headquarters continued on the floor of the Commons. On 9 April 1884, during consideration of the Survey's annual grant, former prime minister Alexander Mackenzie told the House that 'the examination was of a very peculiar nature' and that it would 'be very difficult for the director to carry on the business unless he has the formal support of the Government behind him.'[63] Macdonald concurred with these sentiments, admitting that 'this kind of thing cannot go on. Insubordination is a great vice in any Department, and especially in one of that kind, where science or the zealous application of science or the knowledge of the individual, is absolutely required to make the work of the Department of any service.'[64] Outside observers, such as Principal Dawson of McGill, were equally disturbed by what had transpired during the parliamentary investigation. In a candid letter to the prime minister, a 'much grieved' Dawson described the committee as 'the mouthpiece of Dr. Bell' and warned that failure to defend Selwyn 'while likely to be most injurious to the Survey and to Canadian Science generally, will certainly involve the Government in further trouble.'[65]

An angry Selwyn, even though he knew that he had the support of the government, viewed the committee hearings as nothing short of 'organized conspiracy'[66] and demanded an impartial investigation into the allegations that Bell in particular had made. The government, however, was not foolish enough to sanction another round of public mud-slinging and simply demanded that Dr Bell privately account for his actions. This he did in a plaintive, seven-page letter to the minister of the interior in which he claimed to be a 'sociable, courteous, affable' worker who had been unjustly persecuted by Selwyn. 'It is my misfortune – not my fault – that I have been assailed by him,' an unrepentant Bell argued. 'I could not prevent him from doing this and cannot be blamed for his actions. On the contrary, I think I deserve sympathy for what I have suffered.'[67] There the matter rested, but the rancour generated by the inquiry remained for several years. 'What has struck me this time more than ever,' Otto Klotz, a topographical surveyor and future director of the Dominion Observatory, recorded in his diary after a visit to Ottawa in February 1886, 'is the extreme jealousy that exists between the different departments and between officers of the same department – Dr. Bell called Selwyn the director "a pig-headed stubborn old beggar," the botanist's (Macoun's) work useless – Prof. Macoun told me the geologists' (Topographical) surveyors are unnecessary, that we (land surveyors) should do all that.'[68]

The select committee's report, meanwhile, was a scathing indictment of the Survey. When measured against the yardstick of practical results, the department's operations were found wanting; in fact, it was repeatedly pointed out during the hearings that the Survey's overall performance had slipped since the days of its first director, Sir William Logan. There were the costly field operations that resulted in little more than 'descriptive representations of the surface of the country,' the dated reports that had been 'so seriously delayed in publication as to render them practically useless,' and the 'unpleasant relations between Selwyn and members of his staff.' The committee's greatest concern, however, was the lack of attention given to Canada's economic mineral resources:

The communication to the public of general information, as to the probable extent and chemical characteristics of recognized mineral deposits, and their availability and adaptability to the commercial uses of this and other countries, is certainly a legitimate field for the attention of our Geological Survey, and would tend more to the material prosperity of the country ... than the purely scientific researches so much indulged in, which seem devoted rather to upsetting preconceived theories of antecedent or rival scientists.[69]

To rectify this situation, the committee sarcastically suggested that the travel descriptions, anecdotes and sketches 'while all entertaining ...

certainly should occupy no portion of the Published Reports.'[70] Instead, it recommended that field reports be published quickly and separately, that the gathering of practical information take precedence over pure scientific investigations, and (most important from Macoun's perspective) that 'field operations be confined to subjects more closely allied practically and scientifically to a Geological Survey.'[71]

Dr Selwyn found it difficult to take the investigation seriously because of the way in which his reputation had been mauled during the select committee's hearings. In a letter to a friend, mining engineer Hamilton Merritt, he referred to the committee members as 'parliamentary wiseacres' and the report as 'a tissue of misrepresentation and falsehood ... not worth the paper it is printed on.'[72] His comments to Macdonald were more tempered though equally negative. He considered the recommendations for a more systematic organization and management of the Survey as interfering, uninformed, and misdirected, particularly the proposal that its work be more closely confined to geological subjects. He reminded Sir John that the Survey was 'a geological and natural history survey according to the act.'[73] The inquiry had nonetheless demonstrated that there was a strong feeling that the public was not receiving its money's worth from the Survey – a point that the opposition continued to raise. During the 12 July 1885 Supply vote in the Commons, Edward Holton wondered why the government had failed to act on the committee's recommendations: '[I]t is generally well known that the Geological Survey is in a deplorable, inefficient condition, and that the value of what it is accomplishing for the country from a useful and practical point of view is ... almost nil.'[74] Sir Richard Cartwright, the respected former Liberal finance minister, supported Holton's assertion, although in a more conciliatory tone. 'No doubt, rightly or wrongly, the public opinion is that we have had very little value indeed from the expenditure in the Geological Survey,' he reasoned, 'and it is high time steps should be taken to make this vote more valuable to the people of Canada, who have to pay for it.'[75] This general desire for practical results, Selwyn wisely realized, was not something that would simply go away, and he began to emphasize the Survey's concern with mining activity in his annual reports.[76]

Following his appearance before the committee, Macoun departed for the Lake Superior wilderness in June 1884 to take advantage of the railway construction in the area. Despite what he had indicated to

Dawson during the hearings, however, the field trip was specifically designed to secure botanical specimens. With his younger son, William, as his assistant, he collected along the eastern shore of Lake Nipigon, where a fierce three-day storm forced the pair to take refuge amongst the small group of islands that bear Macoun's name today.[77] They then worked their way along the proposed rail line one hundred miles eastward towards Port Arthur. At one of the construction camps along the way, Macoun, learning of the disruption caused by whisky pedlars, pretended he was a magistrate, confiscated all the alcohol, and took the pedlars prisoners. He repeated the same procedure farther down the line at Michipicoten and then quickly caught a steamer homeward before the real magistrate arrived.[78]

The Professor had no sooner arrived back in Ottawa than he was informed by Dr Selwyn that, at the special request of William Van Horne, general manager of the Canadian Pacific Railway, he would be returning westward by rail with some of the visiting members of the British Association for the Advancement of Science. At the invitation of the Royal Society of Canada and with the prospect of having a large part of the travel costs covered by the Canadian government, the British Association decided to convene its 1884 meeting at Montreal. Begun in 1831, these annual gatherings were designed to enlarge public understanding and appreciation of science and stimulate new research. The Macdonald administration, however, had subsidized this first meeting of the association on Canadian soil in order to advertise the young Dominion's potential as a trade partner and home for Britain's excess population. It was not overly concerned with the encouragement of Canadian science and subsequently balked at funding the various research projects that were launched at the conference sessions.[79]

As part of its promotional strategy, Ottawa, in co-operation with the CPR, treated 150 association members to a free excursion to the Rockies following the meeting. The professor had thus barely time to unpack his specimens from Superior before he was on his way back to Port Arthur, this time as Canadian host to some of the foremost British scientists and interested amateurs of the period. The British Association tour of western Canada was something of a triumphant return for Macoun. The very route that the excursion train followed across the prairies was through the area he had so enthusiastically endorsed only a few years earlier. Macoun was consequently in his element and, as anticipated by Van Horne, freely shared his wide knowledge of the region – by example, when necessary. At the summit of the Kicking Horse Pass, in a scene reminiscent of his days on the Fleming expedition, he led an unsuspecting visiting botanist on a merry chase up Cathedral Mountain as far as his shoes would carry him. A geological party under Selwyn, meanwhile, escaped near disaster when the Survey director's hammering at the entrance of a railway tunnel caused a rock fall. This incident only added to the excitement of the excursion which included a visit with the Blackfoot chief, Crowfoot, at Calgary and a tour of the Bell model farm at Indian Head. Macoun's running commentary also had its impact. Agronomist J.P. Sheldon, in his published account of the tour, remarked, 'Nous devons attribuer une grand du succès et du plaisir de notre voyage à la présence du Professor Macoun ... qui ne laissa pas de nous renseigner.'[80] Other Association members were just as appreciative. 'You added so much to my information about Canada and in our conversations corrected so many of my first and incorrect impressions,' wrote N.H. Martin, 'that I shall ever be your debtor in the matter.'[81]

His botanical interest piqued by the few days he had spent in the Rockies with the British Association, Macoun decided to undertake a more comprehensive survey of the Selkirk Range the following field season. By the spring of 1885, however, rising costs had virtually exhausted the Survey grant for the 1884–5 fiscal year, prompting the director to release Macoun, as well as two other permanent and all temporary employees, as of 30 April. Dr Selwyn did not believe that these men were expendable, but was gambling that the threat of their dismissal would result in a larger appropriation. He was right. When the affected Survey employees took their case to members of Parliament, the additional funds were voted.[82] Macoun then left for British Columbia in late May and spent a memorable summer collecting among the Selkirks near Rogers' Pass.

When Macoun returned to Ottawa that fall, he found the Survey staff busy preparing displays for the Colonial and Indian Exhibition to be held in South Kensington, London, England, the following summer. Conceived by the Prince of Wales in the spirit of the famous 1851 Great Exhibition at the Crystal Palace organized by his late father, the exhibition was to be an international demonstration of the natural resources, manufactured goods, and products of the British Empire. Naturally Macoun wanted to go. Whiteaves, however, was already slated to represent the Survey's natural history section, and he called upon the Professor to assemble a representative collection of Canadian woods, as well as plants, similar to those that he had prepared for earlier international exhibitions. Macoun went about this work dutifully, making a special trip to southwestern Ontario that fall to collect tree samples. At the same time, in what had become a characteristic move to secure special favour or exert pressure, he privately contacted several of the British scientists that he had accompanied to western Canada only a year earlier, calling on them to contact Sir Charles Tupper, now the high commissioner for Canada in the United Kingdom, and impress upon him the importance of Macoun's attendance at the exhibition. The response was overwhelming. Both J.P. Sheldon and Sir Joseph Hooker wrote immediately to Tupper, suggesting not only that Macoun should be available to discuss Canada's botanical features but also that he should have the opportunity to examine the Kew collection for the purposes of his plant catalogue.[83] Others placed themselves at Macoun's disposal. '*Kindly let us know in what direction* we can best aid your object,' W. Fream replied, 'and I need not say how pleased we shall be to do so.'[84] Some, like N.T. Mennell, were simply appalled. 'It would be a most monstrous absurdity if they left you out ... I hope it may prove a false alarm.'[85] So deluged was Sir Charles with representations on the Professor's behalf, including a visit from the eminent Arctic explorer Dr John Rae,[86] that he had little choice but to advise Canadian officials to send Macoun in place of Whiteaves. An expectant Macoun received official word of his appointment in late February 1886. He, along with assistant chemist and lithologist F.D. Adams and assistant mineralogical curator C.W. Willimott, were to assist Selwyn in London.

Despite such questionable tactics, Macoun was particularly well suited to be one of the Canadian representatives at the exhibition. As Dr Selwyn explained to Thomas White, the new minister of the interior: 'there should be gentlemen in attendance, whose personal knowledge of the country and of its natural products and resources, enables them to give intelligent and reliable information and correct replies to the numerous inquiries which will be made on these and other subjects.'[87] Macoun not only met this requirement on the basis of his performance during the 1884 British Association excursion, but, as illustrated by his very presence in England as part of the Canadian delegation, he had a number of influential friends in English scientific circles. These men, in fact, were quite anxious to repay the Professor for what one described as his 'constant kindness in the Great North-West.'[88] He was placed on the distinguished visitors' list and attended a number of official functions and receptions in top hat and frock coat. He also found that he was continually being sought out by British scientists and that, whenever he lectured, his comments were accepted as those of an expert. When, for example, he addressed the Birmingham meeting of the British Association, his statements were supported by 'hosts of *Professors* before our maps who aired themselves to our advantage.'[89] This reception served only to strengthen Macoun's already firm belief in the values of the British connection.

The exhibition itself, according to Tupper's official report, 'displayed Canada's achievements in every department of civilization in such a manner as to astonish many even of our own people ... she asserted, not in words, but in visible deeds, her position as the foremost of the dependencies of Great Britain.'[90] Occupying a 'commanding presence' of some 90,000 square feet, the Canadian court featured sections on the vegetable, animal, and mineral kingdoms, manufactures and industries, education, and the arts. Visitors could view everything from samples of North-West soils in glass tubes and farm implements in motion to landscape painting and a miniature horticultural garden; Macoun's plant collection alone numbered 2,517 different species.[91] The Professor was personally responsible for the forestry section but tended to wander at large and get involved wherever he felt compelled. This behaviour had its benefits. During the erection of the imposing agricultural trophy, he discovered that the potatoes and apples from Anticosti Island were actually wax imitations and had them removed before a public scandal erupted.[92] One of the high points of Macoun's visit was his brief interview with the Prince of Wales during the royal inspection of the exhibition. He also found time to return to Maralin in northern Ireland, to see his former home and visit old family friends.

Despite his absence in England for eight months, as well as the uneasy internal situation at the Survey, Macoun never lost sight of his goal of completing a comprehensive survey of Canada's flora. By the following spring, the glories of the exhibition were behind him and he was back in the field, furiously working up the botany of Vancouver Island, again with the assistance of his son William. This would be the last season that Willie worked with his father in the field, for he soon joined the staff of the Central Experimental Farm as the director's assistant and went on to an illustrious career of his own as Dominion horticulturalist and award-winning apple breeder. Macoun had also realized by now that even with the help of his elder son, Jim, he could not realistically expect to cover every region of Canada as thoroughly as he would have liked for his plant catalogue. He therefore tried to enlist the aid of local botanists, asking them to collect in certain areas or look for particular species. 'Any help I can give you will be gladly available at any time and plants sent to me will be attended to at once,' he responded to a routine inquiry in September 1887. 'I have started on a great work and want all the information possible to make it a success.'[93] Any such co-operation, however, was rather casual and never assumed significant proportions. Given the great physical distances of late nineteenth-century Canada, amateur collectors tended to look to the nearest regional centre for help.[94]

The Professor's dealings with Canadian academic botanists was likewise limited. Far from acting like a magnet in his position as Dominion botanist, he rarely received notes of their findings, let alone specimens, from these men. These university scientists, like Macoun, preferred to send their new or questionable species to acknowledged specialists.[95] The Professor's self-righteous attitude probably did not convince them otherwise. He did, however, believe that he 'had an *honorary right to see* any additions that were made' and was deeply hurt by those who refused to co-operate. 'Here I am labouring away and trying to get our botany put in shape,' he complained to follow botanist Thomas Burgess, 'and instead of being offered assistance I cannot even drag it out of my friends.'[96]

Macoun had no such trouble with foreign specialists. These men looked forward to the arrival of a box of his recent collecting efforts, which not only gave them a clearer idea of the range and composition of Canada's flora but also invariably contained species new to botany. For example, near Victoria in 1887, he discovered, among a number of new species that season,[97] the diminutive Macoun's meadow foam or *Limanthes macounii*, recognized today as one of the rarest plants in the world.[98]

Letters to these specialists made up the bulk of Macoun's correspondence during his tenure at the Survey and resulted, in many cases, in the development of a spirit of professional comradeship. 'Although I have never had the pleasure of personal acquaintance with you,' George Vasey of the u.s. Department of Agriculture closed one of his letters, 'you are one of my oldest and most cherished correspondents and I hope your days of work may be many and long extended.'[99] A similar bond existed between the Professor and one of America's foremost authorities on willows, Michael Schenck Bebb of Rockford, Illinois. In acknowledging Bebb's determination of his willow collection for 1887, Macoun confided, 'I am the party under obligation and for that words cannot convey the estimate of the work done for me by you.' He added,

I can understand your delight in examining what may be new to you as I only feel my earlier sensations when I stand in a new field and see around me new forms and know that every step will add to my enjoyment. This brings up the day of yore when as a young man I tramped the woods alone yet not alone for all around me were new friends that reminded me of even earlier days when I trowel in hand but without knowledge dug up the primroses and violets for my garden in the far off time in Ireland.[100]

Bebb, in turn, replied, '*Obligation is all on my side*. You have for years sent me much material that was *virgin* —save through your generosity quite inaccessible. This has not been only a pleasure for me to study – it has, in not a few instances, been wonderfully instructive – and I simply to give you an enumeration of my determinations in return seems very very small recompense.'[101]

Such praise confirmed Macoun's already strong belief in the value of his work and his capabilities, and he became even more determined to work up the flora of Canada, singlehanded if necessary. 'The day of ... guessing is nearly over,' he informed H.N. Ridley of the British Museum, 'and should I live ten years longer the Canadian forms at any rate will be separated so that my successor may have a clean sheet upon which to make his advances.'[102] It also made all the difficulties, inconveniences, and fatigue incidental to such work seem worthwhile, spurring him on to new collecting heights. 'I must have another summer on the Island [Vancouver], as there certainly are many new things in the north,'[103] he told Vasey. He became obsessed with finding species new to science and with gaining the recognition that went along with their acceptance.

The most immediate outcome of Macoun's wide-ranging collecting efforts, however, was the first volume of his *Catalogue of Canadian Plants*.[104] Published in three parts between 1883 and 1886, the 608-page catalogue drew upon the Professor's extensive field work as well as all existing published records on Canadian flora to provide the synonymy, habitats, and collections for every known species of dicotyledonous plant in Canada. For reviewers in Europe and North America, it was a testimony to Macoun's collecting prowess; in fact, it was judged as going a long way in its expressed purpose of removing the obstacles that stood in the way of a flora of Canada – a project that was still awaiting completion. 'Professor Macoun is to be congratulated on so successful a completion,' the *Botanical Gazette* reported in a typical review, 'and has the wishes of American botanists that the second volume may not be long delayed.'[105] Nor was Macoun himself overlooked. In Great Britain in December 1886, he was made a fellow of the Linnaean Society, while a year later he was unanimously elected a corresponding member of the New York-based Torrey Botanical Club.

That this kind of international recognition should be the outcome of Macoun's first five years at the Survey was rather unexpected. It was a time when there was intense political pressure on the Geological Survey to provide practical scientific information that would facilitate the immediate development of Canada's apparently limitless resource bounty. It was for this very reason that Macoun had been rewarded with a government position. He had used his botanical knowledge to promote the opening and development of the western interior and, upon joining the Survey staff, was generally assumed to be performing similar work. Macoun did much to encourage this impression. In his position as Dominion botanist, he continually spoke of the young Dominion's great destiny and the ways in which a practical understanding of Canada's flora confirmed that destiny. 'It was my knowledge of botany that enabled me to *boom* our northwest as the scientific data I produced were unanswerable,' he advised one of his correspondents. 'It may be yours to boom Newfoundland for depend upon it, a knowledge of its indigenous flora will soon settle the climatic conditions. Were Newfoundland one of our provinces I would bring it before the world in a year.'[106] Yet not once during these five years was Macoun ever specifically dispatched to make an assessment of the agricultural resources of a particular region based on its flora. Instead, as Asa Gray's biographer has argued, he was 'making a reality of the concept of a flora of North America as distinct from a flora of the United States'[107] – exactly the kind of research that Survey critics dismissed as of little use.

As if this pioneering work on the geographical distribution and make-up of Canadian plants were not enough for one man, Macoun's duties were expanded into a full-fledged biological survey of Canada in December 1887, by one stroke of the deputy minister of the interior's pen. Characteristically, the Professor had a hand in this change. In November 1887, after his return from Vancouver Island, Macoun was asked by Robert Ells and Hugh Fletcher of the Survey to introduce them to Thomas White, the recently appointed minister of the interior. After the two geologists had voiced their grievances, the Professor took advantage of the meeting with White to complain about the lack of recognition of his work in Canada and to express his desire to be appointed assistant naturalist.[108] On 13 December, White instructed his deputy, A.M. Burgess, to prepare a memo to the privy council 'creating John Macoun, "Botanist and Assistant Naturalist" with the rank of Assistant Director, in recognition of his long and valuable services to Canada. He has been 15 years in the service. I want it as a Christmas present, so must go in today.'[109] In the draft of the order that was sent to council, however, Burgess scratched out the word 'assistant' before 'naturalist.'[110] Perhaps he thought that Macoun was already fulfilling these duties. In any event, on 27 December 1887, the governor-general appointed Macoun naturalist and assistant director to the Survey with the rank of chief clerk. The Professor had not only survived these turbulent years. He had triumphed.

3

The Search Widens

At the time when John Macoun was promoted from Dominion botanist to Survey naturalist in December 1887, natural science was moving in a different direction from the one he favoured. Knowledge of the wealth and diversity of plant and animal forms was being acquired so rapidly that the day of the all-round scientific investigator with its emphasis on field work was giving way to the era of the university-based specialist who concentrated his energies in one particular field and on the life history of organisms. And with this specialization, the naturalist became increasingly professional in his outlook and started to think of himself as an ornithologist, mammalogist, or herpetologist, not a mere generalist.[1] Macoun's appointment contradicted this trend. In fact, his promotion to Survey naturalist was highly questionable, since he already had his hands full cataloguing Canada's plant life. Macoun, however, had gone after the position, convinced that he could use the same formula that had proved so successful in his botanical work; he simply had to expand his collecting efforts to include all forms of animal life. His promotion and continued emphasis on field work, however, had a decisive impact on the kind of natural science that was carried out under the auspices of the Geological Survey for the next twenty-five years.

Macoun's new position as Survey naturalist brought with it substantially increased duties. Besides continuing his botanical studies, he was now responsible for cataloguing the animal life of the northern half of the North American continent. It was a staggering task, compounded by the fact that the Survey, although responsible for the enumeration of Canada's flora and fauna, was not really in a position to undertake such work. The minuscule financial and manpower resources of the botany division were entirely disproportionate to the immense area requiring examination. All that Macoun could really do under the circumstances was to concentrate on collecting as widely and thoroughly as possible.

This activity had its costs. Working in the field with only the assistance of a part-time helper and occasionally his son, Macoun created the unfortunate impression that it was possible to fulfil his duties with the limited resources available. In reality, however, his field work was little more than a mad scramble: specimens were poorly documented and, in the case of fauna, poorly preserved. Once back in Ottawa, these specimens often sat around for weeks, sometimes months, before they were unpacked. Macoun's concentration on field work and his obsession with securing recognition for finding species new to science also meant that he came to rely heavily on American specialists for the determination and assessment of his zoological collections. Such Canadian-American collaboration had gone on for several decades and was understandable, in light of the fact that Macoun was essentially a botanist. Yet these specialists often took advantage of Macoun's situation and told him not only how to collect but what and where. They used the specimens and accompanying field data that Macoun freely provided to help work up the continent's floral and faunal distribution and composition.

Macoun, for his part, both welcomed and appreciated any assistance that he secured from foreign specialists. Given his phenomenal work-load, he had barely enough time to go over his field season's collecting efforts, let alone work up his various catalogues. Nor did he possess the expertise to determine his difficult specimens – something that was particularly important to Macoun, since he believed that Canada was the home of a number of unique plants and animals. Above all, the Professor was not really interested in doing the kind of work that he turned over to his correspondents because he regarded his own work as more important. He prided himself on being an accomplished field man who, under difficult circumstances, was uncovering Canada's rich and varied natural life. Indeed, from Macoun's perspective, the major problem that emerged out of his first half dozen years as Survey naturalist was not that his large collections were being assessed by American naturalists, but rather that there was inadequate storage and museum space at Survey headquarters to house and display them properly.

The investigation of Canada's fauna was not a virgin field of endeavour for either the Geological Survey or Macoun. Well before the agency had been officially charged with the examination of the Dominion's biological life, geological parties had returned from the field each year laden with natural history items. Robert Bell, George Dawson, and J.B. Tyrrell were particularly active in this regard. In the late 1870s and early 1880s, Dr Bell sent Hudson's Bay Company posts in the northern districts a list of questions about the breeding habits of the local mammal life.[2] He also made noteworthy collections of the flora and fauna of the Hudson Bay region during the 1884 and 1885 field seasons. Dawson and Tyrrell similarly took time from their geological duties to make observations on the distribution of animal life, as well as to collect specimens whenever possible. Tyrrell was in fact able to publish in 1888 a small catalogue on the mammals of Canada based largely upon his five years' work in western Canada.[3] Macoun's rival, J.F. Whiteaves, had also done some faunal work in addition to his palaeontological studies. Before joining the Survey, he had dredged for marine invertebrates in the Gulf of St Lawrence, and in 1886 he had prepared the catalogue for the Department of Marine and Fisheries display at the Colonial and Indian Exhibition.[4]

Professor Macoun, for his part, had always taken a general interest in the various forms of nature. He started to study birds seriously in 1880 during his exploratory work for the federal government in the western interior. Employing the same method he had used when he first started his botanical pursuits, he described in a notebook the birds that he had observed or shot and later compared his notes with descriptions in books. It was a tedious procedure, but as far as Macoun was concerned there was no other way to acquire a practical knowledge of ornithology than in the field. 'That is where,' he told Elliot Coues, one of America's premier ornithologists, 'original work ought largely to be done.'[5] The results of these bird studies were published in *Manitoba and the Great North-West*, along with separate chapters on the mammals, reptiles, fish, and insects of the region. These ornithology notes were also later used by Ernest Thompson Seton for his work on the bird life of Manitoba.[6]

After joining the Survey, Macoun collected not only plants but any other interesting form of animal life that he found in the field, including marine invertebrates. In bringing back these specimens, he was fulfilling the responsibilities of the Survey as outlined in the 1877 Act. It was a duty, however, that the botanist would have undertaken even if it had not been officially prescribed. Imbued with a great love of the natural world, he could not help but marvel at the great wealth and variety of animal life that he came across while botanizing and, since he was already in the field, it took little effort to gather these additional specimens. Sometimes, these chance collections resulted in a significant find, as in 1886, when he discovered a new species of butterfly (*Oeneis macounii*) in the Nipigon district.[7] Such success served to fuel his passion for field work and, by the time he was named survey naturalist, the desire to collect fauna on a full-time basis was 'a fire that has long burned within me.'[8]

Macoun originally wanted his son Jim to assume Whiteaves' duties as zoologist. In a letter to Selwyn written prior to his departure for the Colonial and Indian Exhibition, he observed that the natural history work of the Survey was assuming such proportions that 'it may be necessary to relieve Mr. Whiteaves of the drudgery of working up all the relationships of the birds, beasts and fishes.' This work, Macoun now argued, could be handled by Jim, with some assistance. He would have preferred to assume it himself but, as he told Selwyn, 'I feel that botany is quite enough for me.' Macoun's comments about his work-load aside, he was probably only too aware that Whiteaves, having recently been bumped from the Survey team slated to go to London, would have refused his assistance outright. He consequently decided to go behind Whiteaves' back and try to have his son permanently appointed, on the grounds that such a move would benefit Whiteaves and the Survey. 'Public opinion is fast taking shape in the matter of Natural History,' he warned Selwyn, 'and a short time in the future many will ask what we are doing ... and somebody must answer.'[9]

James Macoun had a good claim for a position on the permanent Survey staff as zoologist. Like his father, he did not confine his field work solely to botany but investigated the local fauna as well. During A.P. Low's 1885 reconnaissance survey of the little-known Lake Mistassini district of Quebec, he collected birds during the spring migration. The following two field seasons, while attached to a topographical party, he made notes on the bird life between Lake Winnipeg and James Bay. Dr Selwyn fully recognized the younger Macoun's abilities and happily forwarded his application for a permanent position to the minister of the interior in January 1887.[10] The matter was abruptly dropped, however, when the Professor was named naturalist later that year.

Just as confusion surrounded Macoun's original appointment in 1882, there also seemed to be some doubt about the botanist's promotion. Except for a note from A.M. Burgess, deputy minister of the interior, in December 1887 informing him of his new position, Macoun had heard nothing further about the matter. The Survey, moreover, was not officially notified about Macoun's new title until 15 March 1888, when Dr Selwyn finally received a copy of the order-in-council.[11] Selwyn was furious. It was not that he was opposed to this kind of work; in fact, in an 1885 memo to the prime minister on Survey reorganization, he had suggested that the agency might be divided into biological and mineralogical sections.[12] What bothered him about Macoun's promotion, however, was that it not only constituted gross political interference – he was not even consulted – but it threatened to aggravate already poor staff relations. Whiteaves, as zoologist and assistant director with specific responsibility for the museum, was understood to be in charge of the Survey's biological work. That was why Macoun had originally asked the minister of the interior for the appointment as assistant naturalist. His promotion to naturalist now brought the entire organization of the Survey's natural history activities into question.

Faced with a *fait accompli*, Selwyn could only ask Macoun to renounce his new title. Yet Macoun, despite his earlier admission that his botanical work was enough for him, would not give up what he regarded as his due reward, and Selwyn had little choice but to delineate the duties of the two rivals. As of March 1889, Macoun was responsible for assembling a representative collection of all the birds and mammals of the Dominion for the Survey museum and specimen cabinets. Whiteaves, meanwhile, would continue to exercise complete control over the arrangement and preservation, as well as the naming and labelling, of all natural history specimens on display in the museum.[13] At best, it was an uneasy truce that would continue to plague the internal operations of the Survey for the next twenty years.

Its primary victim for the time being was Samuel Herring, the Survey's taxidermist since 1883; subject to the instructions of both men, he often found himself caught in the middle.

Macoun's first task as naturalist was to deal with the criticism that had recently been levelled at the Survey's ornithological work by Montague Chamberlain, a New Brunswick businessman and one of two Canadian ornithologists who participated in the founding of the American Ornithologists' Union (AOU) in New York in September 1883. The AOU was a response to the alarming decline of several bird populations in the United States through senseless slaughter, as well as to the need for some national organization to deal with issues such as nomenclature; its creation also reflected the emerging sense of identity among American ornithologists. One of its immediate tasks was the determination of the geographical distribution and seasonal migration of the birds of North America, and it called upon amateur bird watchers throughout the continent for assistance. Chamberlain, whose nickname was 'bulldog,' contacted Macoun in late December 1883 in his new capacity as supervisor of the Canadian section of the AOU committee on migration and geographical distribution and urged him to turn over his bird notes. The professor agreed to co-operate and, over the next few years, he and his son sent Chamberlain regular instalments. It was an unpleasant exercise, however, for Chamberlain was forever questioning the reliability of Macoun's data, while chastising Jim for making amateurish mistakes.[14] Robert Bell, another Survey contributor, received similar treatment, being characterized as knowing 'nothing of the subject' yet imagining 'he knew everything.'[15]

This poor opinion of the Survey's ornithological work, together with some apparent encouragement from Joseph Whiteaves,[16] prompted Chamberlain to call upon Dr Selwyn in the fall of 1887 and argue the need for a staff ornithologist. The director was not convinced. As Chamberlain later told Macoun: 'I attempted to show to the Director that the Survey might without much additional expense do a large amount of valuable work for ornithology. He told me that all that could be learned about our birds was already known to science, called my statements "rubbish" and ended the interview by turning his back on me and proceeding with his writing.'[17] Somewhat miffed, Chamberlain decided to pursue the question with the minister of the interior at roughly the same time that Macoun had gone to him for recognition of his botanical work. He had no illusions about his chances – a feeling that was subsequently confirmed when Selwyn responded to his formal application to the minister. At least this time, Selwyn showed more civility. 'While not in the least doubting your fitness for the position, nor undervaluing the importance of ornithological study and investigation,' he explained, 'I consider the position and circumstances of the Survey and the Museum connected with it, make it at present inexpedient and unnecessary to make a special appointment.'[18]

The damage to Chamberlain's self-esteem, however, had already been done. Resuming work on his unfinished *Catalogue of Canadian Birds* manuscript, he turned to the preface with a vengeance. After briefly outlining the object and deficiencies of the catalogue, he emphasized the need for further detailed investigation, taking direct aim on Selwyn's position. 'I am quite aware,' he observed, 'that this opinion regarding the narrow limits of our knowledge of Canadian birds is opposed to that held by some of the leading scientific men in the Dominion who consider that all that can be learned about our fauna is now known to science.'[19] Chamberlain then went on to quote extensively from private letters he had received from leading American ornithologists deploring the lack of ornithological work by the Survey. Professor J.A. Allen of the American Museum of Natural History in New York and president of the AOU had written, 'I have long watched with interest the reports of the Canadian Survey, and have been disappointed to find the natural history portion of the work receiving so small a share of attention, where the field is so inviting and as yet so little worked.'[20] The other letter excerpts were equally damaging.

Such comments were not altogether unfair. Even though the various field parties collected natural history specimens, their geological duties took precedence. As for the other staff members connected with the natural history work of the Survey, Herring was engaged full time in taxidermy, while Whiteaves was often called upon to assume the director's administrative duties when Selwyn was absent. The Survey's bird collection was consequently quite deficient in comparison to the holdings of other institutions and private individuals, and the specimens for the Canadian display at the Colonial and Indian Exhibition had to be purchased or borrowed from Ontario collectors.[21] In fact, the only serious attention the birds' skins received at the Survey was from dermestid insects in the specimen drawers. Chamberlain's comments were therefore quite beneficial in that they eased Selwyn's misgivings about Macoun's promotion and prodded him to order the immediate sorting, cataloguing, and labelling of the Survey's bird collection. At the same time, they put him on the defensive and, instead of seeing that a full-time ornithologist was added to the Survey staff, Selwyn expected Macoun in his new capacity as Survey naturalist, and to a lesser extent his son, to handle this work. This decision proved short-sighted in the long run.[22]

Despite the immense area requiring investigation and the limited staff resources at his disposal, Macoun did not wince at the prospect of the work ahead, for, as he confided to one of his botanical correspondents, 'By sheer force of will and without any assistance I have mastered all difficulties placed in my way and am now Naturalist to the Government ... What I want is life with working power ... as work to me is living.'[23] He was greatly distressed, however, by Chamberlain's criticism, interpreting it as an attack on his appointment. 'Mr. Chamberlain is making a dead set on us,' he grumbled to another New Brunswick ornithologist. 'You would scarcely know by his book that he was aware of my bird knowledge during the last ten years. I cannot understand why he makes out we have done nothing.'[24] The professor felt that he had to prove himself and advised Chamberlain in early April 1888, 'I assure you that no matter how much reason you had for thinking that we were doing nothing in the past, you will not be able to say so in the future as I am going to shake up the dry bones.'[25] Chamberlain, for his part, accepted defeat gracefully. 'I see that you have a lecture in store for me,' he responded in a congratulatory letter, 'if any statements that I have made are not correct, I am ready to withdraw them as publicly as they have been made. Whatever comes to pass, you and I need not quarrel on it.'[26]

Macoun and Selwyn were not so forgiving and, in a somewhat spiteful move, it was decided that the new Survey naturalist would produce a bird catalogue modelled along the same lines as his successful plant publications. Before taking up the project, however, Macoun left on a prearranged botanical survey of Prince Edward Island, taking along his wife and two youngest daughters. It was thus not until November 1888 that Macoun started work on the bird manuscript with every intention of completing it by the following winter. He also began to make preparations for the collecting of faunal life and actively sought the advice of some of the foremost American authorities in the field. He approached ornithologists Elliot Coues and J.A. Allen about bird sightings, B. Goode of the u.s. National Museum about mammal preparation, and B.W. Evermann, a future officer with the u.s. Fish Commission, about collecting fish. In providing this information, Evermann encouraged Macoun to 'do for the Canadian fishes what you have admirably done for the plants.'[27] The Professor's most important American contact, however, was C. Hart Merriam, chief of the Division of Economic Ornithology and Mammalogy in the u.s. Department of Agriculture, and roughly Macoun's counterpart.

Merriam had first met Macoun through Chamberlain at the 1884 Montreal meeting of the British Association for the Advancement of Science (BAAS) and was so impressed with the botanist's enthusiasm that he took advantage of his proximity to Ottawa to visit the Geological Survey headquarters. Merriam later informed Macoun that his time in Ottawa was 'one of the pleasantest episodes of my life' and that he 'had learned enough to well repay me for the loss of time.'[28] In the same letter, however, he could not resist the temptation to take issue with the Professor's views on the matter of creation. A committed Darwinian who believed that all terms of classification such as variety, species, and genus were arbitrary, Merriam lectured: '... what you are pleased to call "*types of creation*" ... have a harsh unnatural sound to my ear ... It implies a state of things which, in my opinion, never did exist and never can exist; and its employment should be restricted to allusions to the dark ages before scientific thought had gained sufficient strength to break the cruel fetters that for many centuries barred the progress of the human race, and retarded the development of the human intellect.' He then closed on a more complimentary note, describing his new Canadian acquaintance as 'a man of extraordinary experience in the field, of unusual powers of observation, of exceptional familiarity with the forms of life scattered over half a continent.' Merriam had clearly sensed during his brief visit to Canada the importance of flattering Macoun's ego.

Merriam had good reason not to offend Macoun: he needed him. Operating on the somewhat erroneous assumption that temperature was the prime determinant of faunal distribution, Merriam was busily engaged in a comprehensive survey of the biogeography of North America.[29] Given the sheer size of the project, any information on species range and habitat or field specimens that the Survey could provide was therefore most welcome, and Merriam actively encouraged Macoun and his son to avail themselves of his services. This they had been doing since 1885, sending Merriam bird and mammal specimens, along with their field notes. Merriam, in turn, did much more than identify and assess the significance of these collections. Since it was to his benefit to improve their field methods, he regularly sent them technical information, ranging from the proper method of trapping small mammals and preparing skins to forms for noting the arrival of birds during spring migration.[30]

For Macoun to seek the assistance of American naturalists in his new capacity as Survey naturalist was only logical in light of his botanical work. Almost from the time that he first became interested in plants, he had sent many of his collections to American specialists – a trend that increased the more serious he became. This reliance on the botanical community south of the border was not simply a consequence of the growing American expertise in North American botany. The professor had become increasingly disenchanted with British botanists, who sometimes failed to acknowledge receipt of the specimens, let alone name them.[31] It was much easier to send his collections south; his specimens would not only receive attention from botanists interested in the continent-wide distribution of species but the exchange period would be considerably shorter. Turning to American scientists also made good sense in that Macoun's field work was largely confined to the southern portions of Canada where he was more likely to find species similar to those in the United States. In fact, he would often ask his American correspondent for samples of particular species likely to occur north of the border.

From a strictly realistic point of view, moreover, Macoun clearly needed the knowledge and expertise of zoologists such as Merriam. Apart from his work on plants and birds, his natural history studies had essentially been based on what he chanced upon in the field. He was little more than an enthusiastic amateur and generally incapable of working up his collections himself, even if he had had the time to spend on them. Since collecting animal life was not the same as collecting plants, it was also vital for Macoun to learn the best field techniques, particularly now that he had to collect as thoroughly and widely as possible during his limited time in any one area. Then again, the

Professor rarely admitted his shortcomings. In response to Chamberlain's criticism of the ornithological work of the Survey, he flatly told Merriam in a typically exaggerated statement, 'I have done more ornithological work in Canada than any other man and ought to know something about our birds but you people are ignorant of this and cannot be blamed.'[32]

Despite the anti-American bias of nineteenth-century Canadian politics, this Canadian-American collaboration in biological research was nothing new. Between 1859 and 1862, Robert Kennicott, a field worker with Washington's prestigious Smithsonian Institution, had toured Rupert's Land and converted a number of Hudson's Bay Company men into avid collectors and correspondents. These fur traders were regularly contacted over the next decade by Spencer Fuller Baird, the Smithsonian's assistant secretary, who operated the Institution's exploration program 'in hemispheric rather than national terms.'[33] By means of flattery, promises of formal recognition, or parcels of books and alcohol, Baird ensured a steady flow of natural history material from the fur traders: in 1863 alone, one and a half tons of specimens were shipped to the Smithsonian. Roderick MacFarlane of Fort Anderson, a collector whom Macoun later met during his exploits in the Peace River country in the 1870s, was particularly prolific and sent more than 5,000 items to Washington.[34]

This continental connection was also exemplified in the professional contact that Canadian scientists had with their American neighbours. Although the BAAS occasionally held its meetings in Canada and boasted such prominent Canadian members as Prime Minister Macdonald, its counterpart, the American Association for the Advancement of Science (AAAS), had more success in attracting and retaining Canadian members during the latter half of the nineteenth century.[35] Principal Dawson of McGill University, for example, was president of the AAAS at its 1882 meeting in Montreal. When the veneer of Anglo-Canadian identity and imperial sentiment, moreover, is stripped away, the relations between the British and Canadian scientific communities were at times testy. In 1882, a London *Times* article ridiculed the decision of the BAAS to hold its 1884 meeting among the simple-minded colonials of Montreal. 'They can have no serious purpose of holding a scientific meeting in Montreal,' the *Times* commented. 'Their wish must be to have an agreeable outing, to be looked up to with blank wonderment ... If they talk nonsense, they will be listened to all the same, and with the same degree of intelligent appreciation.'[36] R.A. Proctor, editor of *Knowledge* magazine, countered by arguing that the proposed BAAS meeting would be a serious mistake since the AAAS had just met in Montreal. 'I know what American tastes are in matters scientific,' Proctor cautioned, 'how much they prefer fresh to dried food in science, and I know that the kind of food purveyed, for example, at Southampton this year, would emphatically not suit American tastes, whether in the United States or Canada.'[37] The world of Canadian science thus did not simply revolve around Great Britain; the idea of imperial science was 'more rhetorical fancy than anything concrete.'[38] Canadian scientists like Macoun also identified with their American colleagues and would probably not have supported the formation of the Royal Society of Canada in 1882 had it not been for the appeal to Canadian nationalism.[39]

With the support and encouragement of American natural scientists, Macoun began to tackle his new duties as Survey naturalist, in particular the bird catalogue. The single-minded intensity with which he went about this work, however, quickly landed him in bed on 15 December 1888 with a severe attack of angina. For one with Macoun's drive, the illness was only a temporary set-back, and he was back at the office working half days within a month. But while convalescing, he had given more thought to his work-load and decided that it would be best to try to limit the amount of office work in favour of concentrating almost exclusively on his field activities. With this in mind, he approached Thomas McIlwraith, a Scottish-born Hamilton coal merchant and one of Canada's leading ornithologists, about the possibility of jointly preparing a field guide to Canadian birds. Macoun would contribute range and habitat information, while McIlwraith, the other Canadian co-founder of the AOU and the author of *The Birds of Ontario* (Hamilton 1885), would provide the descriptions. 'By doing this,' Macoun reasoned, still smarting from Chamberlain's barbs, 'you will set us right before the country.'[40] There was no need to hurry the book either, for during Macoun's illness, his son had gone over the Survey's bird and mammal collections and found that the area west of Lake Superior, especially the Rocky Mountain region, was poorly represented. Macoun therefore suggested that he would spend the 1889 field season in British Columbia.

McIlwraith, in response, thought the idea a splendid one. Through-out the 1880s, he had continuously resisted turning over his Ontario bird notes to Chamberlain in the expectation of preparing his own publication on the birds of Canada.[41] Macoun's proposal now re-kindled this hope so much that he wanted the Professor to stay at home and get to work on the bird book immediately: 'hitherto you have done too much work ... and left the supervision to others ... I think your time is too valuable to be spent collecting specimens of any kind which can be done by others at less of a sacrifice. You think I have done pretty well collecting without going from home at all?'[42] Macoun, however, had already made up his mind to work up the birds in the field and was not to be deterred from this self-imposed assignment – even at the expense of his health. As he told McIlwraith, 'You have collected a great number of birds, but I *want* to collect them myself so I must know *where* the birds live and *how* they live as I do the plants.'[43]

Although he had regularly collected faunal specimens in the past, Macoun considered the 1889 field season as his first official outing as Survey naturalist. And unlike his earlier work for the Survey, he was accompanied not only by his son but also by a new field assistant, William Spreadborough of Bracebridge, Ontario. Spreadborough's very presence was remarkable given the professor's determination to try to do most of the field work himself. 'I have not the slightest faith in subordinates in carrying out their instructions,' he had bluntly told the 1884 Select Committee of Inquiry into the Geological Survey.[44] His expanded work-load and recent illness, however, seemed to have softened his resistance to field help. Besides, Spreadborough came highly recommended. The previous field season, Jim Macoun had served as naturalist to the Thomas Fawcett survey of the waterways between Lesser Slave Lake and the Manitoba lakes region. Spread-borough, who had recently lost his young bride during childbirth, had signed on as cook, but soon showed a marked aptitude for natural history, stemming from his boyhood days on the family farm on the south branch of the Muskoka River. One day he mentioned to Jim that he had heard a 'firebird' or scarlet tanager about camp. Jim gave him his shotgun and, to his surprise, the camp hand returned with a fine specimen of the relatively unfamiliar western tanager.[45] Jim thereupon decided to teach Spreadborough how to prepare specimen skins, and was so pleased with his rapid progress that he subsequently convinced the senior Macoun to take Spreadborough on for the 1889 field season. From that time forward, for the next thirty years, Spreadborough was a regular member of the Macoun or other Survey parties. In fact, according to future Survey ornithologist Percy Taverner, Jim Macoun 'practically grew up, in summer time at least, under the eyes and more or less the protection of William.'[46] Usually the advance man in the field, Spreadborough was generally free to collect wherever he thought best and often spent several weeks alone in relatively isolated areas. He also continued to be in charge of camp operations. 'His bannocks were things to make one's mouth water,' wrote Taverner shortly after Spreadborough's death, 'and he knew how to make a most unpromising camp comfortable. Gathered around the fire chilly nights his running comments on things and tales of camp experiences were better than most books.'[47] Despite this valued field assistance, Spread-borough remained largely unknown outside the small circle of Survey employees engaged in natural history work.

Macoun's first official season as Survey naturalist was an auspicious one. Starting from Vancouver on 4 April 1889, the trio slowly worked their way eastward along four hundred miles of CPR line to the Gold Range. By day, they cast their net over the surrounding countryside, gathering in all forms of flora and fauna they happened upon. At night, their catch was put up. Plants were dried and pressed in felt blotters leaned against stakes around the campfire, birds and mammals were skinned and sprinkled with arsenic, fish and reptiles were plunked in alcohol-filled jars, and insects were meticulously placed in gauze cases. For anyone who might have stumbled upon their camp, the scene could easily have been mistaken for a kind of devil's workshop. Their labours that summer were richly rewarded: 15,000 plant specimens, 435 bird and mammal skins, 100 reptiles and fish, and quantitites of insects.[48] It was just the kind of outing Macoun needed, for as he wrote to Sereno Watson upon his return to Ottawa, 'My health is completely restored so that I can go to work again with my old time zest.'[49]

That winter, in working over his summer's findings, Macoun carried on an extensive correspondence with Dr Merriam.[50] He not only sent him many of the bird and mammal skins from British Columbia, but all the reptiles and fish he had collected over the past few years. Merriam passed these specimens along to doctors T.H. Bean and Leonhard

Stejneger of the Smithsonian who named them in return for keeping all duplicates. Macoun also continued to seek collecting and preservation advice. He called upon Merriam to send him forms for noting the arrival of birds and to outline the best means of keeping insects from entering and destroying specimen skins. He also asked him to approach Bean and Stejneger for the names of those species considered rare. Merriam attended to these various demands as best he could, since he counted upon Macoun for much of his information about the geographical distribution of Canadian faunal life. Unfortunately, the Professor came to be taken for granted. In writing to his Canadian colleague before the start of the 1890 field season, Merriam casually remarked, 'Of course you will make collections of mice, shrew, and other small mammals, and obtain as much information as possible concerning the distribution of all species.'[51] Macoun did not mind this kind of treatment, for Merriam and his associates had identified a new species of fish (*Chauliodus macounii*) and what was initially thought to be a new variety of chickadee from among his 1889 collections. He was convinced that his strategy of concentrating on the field end of operations was the correct one, particularly when there seemed to be so much more to learn about Canadian fauna. 'It is quite evident we have plenty to do north of the line,' he confessed to Merriam in a rare moment: 'even in birds there is something new to be had. I hope to add some more data to that of last year and of course I will report it to you.'[52]

Anxious to push on with the bird catalogue, Macoun had Spreadborough remain behind in British Columbia over the winter of 1889–90, collecting birds as well as small mammals in the Victoria area. He then had him proceed eastward in the early spring to the Kootenay district to intercept migrating birds as they passed up the Columbia River. The Professor joined Spreadborough in late May but quickly found that they could not keep up with the collecting season – in fact, by the time Macoun had reached Revelstoke, Spreadborough had already procured some 160 bird skins. Son Jim was consequently summoned from Ottawa and as they had done the previous summer, the threesome carried out an extensive survey of the natural life within the great bend of the Columbia.

The following season, Macoun intended to spend the summer closer to home. On 3 June 1891, he left for western Ontario with H.N.

Topley, photographer to the Department of the Interior to take pictures of trees for the Canadian exhibit at the 1893 Chicago World's Fair. This trip was cut short, however, at the request of the new minister of the interior, Edgar Dewdney. He had other plans for the Survey naturalist. In 1887, the Macdonald government had established Canada's first national park, Rocky Mountains Park, at Banff in the Bow River Valley along the southeastern slope of the rockies. Far from being designed as a wildlife and forest reserve, the park had been created to promote international tourism in this otherwise 'useless' section of the CPR main line. The scenery and hot springs were being protected from private developers so that the government, in co-operation with the railway, could exploit them. Dewdney fully endorsed this rationale for the park's existence and, in an effort to encourage other forms of economic activity, proposed a park museum illustrating of the resources of the area. When he got in touch with Selwyn about the matter and found that the Survey did not have material suitable for the proposed museum, it was decided to dispatch the Professor to make a representative collection of the flora and fauna of the area.[53] This assignment nicely dovetailed with Macoun's own work and, with Spreadborough once again at his side, he spent the better part of the summer exploring the Bow River Valley along the southeastern slope of the Rockies, followed by a short collecting spree at Indian Head, Assiniboia, on his way back home. It was another successful season – 250 bird specimens alone – that was marred only by an unpleasant run-in with a local North-West Mounted Police inspector over the naturalist's failure to secure a permit for the preserving alcohol that he had taken to Banff. When Dewdney later sought Macoun's opinion on the qualifications for the park museum curator, he frankly advised his political superior, 'get a man who is willing to make himself something and who will do the country good when speaking intelligently about it. A political bloke should be the last one for that place.'[54]

Because the collecting prospects at Indian Head seemed so encouraging, Spreadborough was stationed there in the spring of 1892 and quickly discovered that the area served as one of the major breeding grounds for North American waterfowl; within two months, he obtained almost 190 bird skins. He also became the surrogate mother of two young owls and when not out collecting, was kept busy trying to satisfy their appetites.[55] From Indian Head, Spreadborough continued his field work at Port Arthur and Nipigon before moving onto Lake Erie where the Professor was busy gathering freshwater fish for a possible catalogue on the subject. The next year, intent on making his work on Canadian birds as complete as possible, Macoun passed up the chance to go to the Chicago World's Fair in favour of spending the summer on Vancouver Island. As had now become the custom, Spreadborough was once again sent ahead, this time to study the avifauna inhabiting the islands in the Strait of Georgia. When the Professor arrived, the pair dredged for marine life off Cape Lazo near Comox and then in Sooke Inlet at the southern tip of the island. Macoun then took a month's holiday, rejoining his wife and youngest daughter, who had come west with him to visit his eldest daughter, Clara, who had recently married surveyor A.O. Wheeler and was living at Victoria.

Botany, in the meantime, had not been forgotten. Whenever Macoun took to the field, he continued to take along his plant basket and press. Nor did he do anything, once he had been promoted to Survey naturalist, to try to reduce his botanical commitments. Anxious to publicize his accomplishments in the field, he sent complimentary copies of the first volume of his plant catalogue to dozens of American botanists, as well as placing a note in the *Botanical Gazette* offering to collect plants for interested individuals.[56] He also continued to publish in the area. By 1890, the second volume of the *Catalogue of Canadian Plants*[57] had been completed, bringing the total number of parts issued to date to five. Like the first volume, it was favourably received by his peers on both sides of the Atlantic as a valuable addition to North American botanical literature. The *Midland Naturalist* stated that 'the treatment throughout is excellent and characteristic of the scientific acumen and indefatigable zeal of the author,' while the *Bulletin of the Torrey Botanical Club* suggested that Macoun was 'contributing more at the present time to our knowledge of North American botany than anyone else.'[58]

The expansion of Macoun's duties did, however, have an impact on his botanical studies, in that he increasingly concentrated his energies on the discovery of species new to science. He no longer was able to undertake the more comprehensive botanical surveys that had characterized his work in the 1870s and 1880s. His enormous work-load also meant that he became more than ever dependent on specialists for the determination of his botanical collections. 'With not only botany but the general history of the entire Dominion to attend to,' Jim wrote to Dr Nathaniel Lord Britton of Columbia College, New York, 'we cannot hope for a long time to do any special work and are thankful to have our material examined by you.'[59] The most important consequence of Macoun's promotion was that he turned a steadily increasing amount of the routine botanical work over to his son – a reversal of the roles that the Professor had earlier outlined to Selwyn. Still a temporary employee, Jim provided yeoman service. During 1888, the same year in which he spent seven months in the Hudson Bay region, he received 1,500 specimens from amateur collectors for determination, mounted and placed 3,015 sheets of specimens in the herbarium, and sent 2,152 duplicates to various institutions and private collectors in Canada, the United States, and Great Britain.[60] The following year, he placed 4,406 sheets of specimens in the herbarium, while distributed 5,960 specimens. The volume handled seems all the more incredible since Jim was often called away from his herbarium duties to help his father with one of his various projects. In February 1890, for example, Macoun was so busy getting ready for the field that he turned his correspondence over to his son with the order to 'make short work of it.'[61]

Macoun's collecting record during his first few years as Survey naturalist was phenomenal, not only because of the area and terrain covered but also in terms of the wide variety and amount of natural history material collected. The epitome of the field naturalist, he had developed a keen knowledge of the natural environment. He also had the remarkable ability to spot new species anywhere he went in the field – a skill that he prided himself on. 'I am an old hand at this thing,' he bragged to American botanist G.N. Best; 'I can spot a new form without an effort.'[62] This confidence in his own field abilities, together with his desire to handle the field work himself, caused Macoun to turn down a steady stream of applications to work with him. In fact, whereas others might have welcomed assistance in the field, he was quite happy to have only Spreadborough and occasionally his son at his side. 'These two men and myself,' he told the minister of the interior in reference to an application for summer work with the Survey, 'can do all the work as we know exactly what we want ... I want no more.'[63]

There were also good economic reasons for not wanting a larger field party. During this period, the Survey was starved for funds; although the cost of running the department had increased annually, the appropriation had remained at its 1885 level of $78,500. Dr Selwyn consequently faced the near impossible task of trying to hold the line on expenses while carrying out the normal functions of the Survey, and was often, as Macoun once observed, in 'a very crusty state.'[64] This severe financial situation adversely affected the timing of Survey field operations, a particularly crucial factor for natural history work. In 1893, for example, despite Macoun's pleading that it was essential to get to the field early before spring migration, Selwyn refused to allow Spreadborough to start for Vancouver Island until early April.[65] The Survey's financial state also placed certain restraints on the field operations themselves. Although Spreadborough was paid per day plus room and board, it was understood that whenever he went ahead in the spring he would travel second class and live as cheaply as possible. Macoun himself practised similar economy in the field, usually camping out even when near a major town. In the interests of using his annual budget to the best advantage, then, the Professor refused any additional field assistance; the only time he took extra help to the field was in 1893 when the minister of the interior was trying to find work for a friend's son.[66]

Macoun's attitude towards field work and his refusal to lobby for additional personnel had its costs. In the first place, it created the false impression that he could uphold the Survey's natural history responsibilities with only limited assistance – an impression that was confirmed by his constant trumpeting of his field accomplishments. But with only three men engaged in the field work of the botany division on a regular basis, a number of practical problems abounded. Each time Macoun, his son, and Spreadborough took to the field, the overriding concern was to collect as much and as quickly as possible. As a consequence, specimens were not gone over until at night in camp at which time they were tagged on the basis of memory. In a few instances, specimens were not labelled for several weeks, such as in May 1892, when an overwhelmed Spreadborough reported to the Professor, 'I have over a hundred and twenty skins and not one taged [sic] as I have been waiting for tags.'[67] There were also problems with the labels. Often they gave only a general location – where the collecting party was staying, not where the specimen had been found. Nor was it unusual to find specimens from two or more localities mixed together in the same package or specimen dates that did not correspond with Macoun's travels.[68] The quality of the specimens also suffered from the attempt to do so much in the field. Acknowledging the receipt of a vole in 1889, Merriam commented, 'The remains look as though the skull had been placed on an anvil and hammered for an hour or two. Still I have been able to pick out enough separate teeth to ascertain the animal.'[69] He was still complaining five years later. In trying to determine a batch of mice sent by Macoun, Merriam reported, 'The genus is a particularly difficult one to study under any circumstances and the task is rendered more perplexing with material such as this, unaccompanied by field measurements and the tails drying up, without being wired and presenting all kinds of conceivable distortions.'[70] Dr Stejneger of the Smithsonian was astonished to find that a reptile that had never been found east of the Rockies was labelled as supposedly coming from Prince Edward Island.[71]

These problems of the field were compounded by the situation at Survey headquarters. There was only one small room devoted to natural activities, and the perpetual question facing Macoun upon his return from the field was where to stash his summer's booty. In his annual report for 1890, the Professor pointed out that the herbarium cases were so full that the specimens were frequently injured by being crushed.[72] Sometimes entire collections were misplaced or the specimens sent to specialists were different from those that Ottawa had retained.[73] Dr Selwyn did not help matters. Criticized in the House of Commons in 1890 for not giving detailed written instructions to Survey parties prior to their departure for the field, he issued a circular of instructions in which he advised among other things: 'All opportunities for collecting specimens of recent Natural History should be made use of when the doing so will not interfere with the main objects of the exploration. Even when collections of these cannot be preserved useful and interesting observations on occurrences and distribution of species can be given.'[74]

So much field data and so many specimens were consequently being collected that it quickly reached the point where there was too much office work for too few hands. The Professor was not deterred, however, but simply turned over more of his routine duties to his poor son, who found it impossible to complete this work before it was time to go to the field again. By April 1891 the office work was so far in arrears that Macoun decided to have Jim remain in Ottawa that summer to clear up the backlog. Three months later, however, his son was named secretary to Survey geologist George Mercer Dawson, one of two British representatives who were being sent to investigate the impact of open-water and land-based hunting on the Behring Sea fur seals in order to determine whether there was any basis to the American charge that Canadian pelagic sealing was destroying the herds. Jim's services were effectively lost for the next two years, as he divided his time between visits to the seal rookeries on the American-owned Pribilof Islands and work on the British-Canadian case before the 1893 Paris Behring Sea Arbitration Tribunal.[75]

With Jim gone, there was a dire need for some form of office assistance. The financial constraints that plagued field operations, however, also affected office work. Macoun was under strict orders from the director to secure only Canadian plants for the herbarium and to trade for, not purchase, them. He was also instructed to give away copies of his plant catalogue only if he received something in return.[76] With Selwyn vigilant about such comparatively small matters, any form of office help at this time was highly unlikely.

Macoun could have helped matters considerably by staying in Ottawa for one or two field seasons. This is something that Selwyn had suggested to him in early 1891 before Macoun was given the assignment to collect material for the Banff museum. The Professor, however, gloried in being in the field close to nature. Whenever he returned to Ottawa after an arduous summer, he usually collected around the capital well into the fall. He was also becoming quite disturbed by the growing emphasis on specialized research in the laboratory away from broad, practical field study. He still believed that it was the naturalist's role to labour diligently in the field and carefully observe, catalogue and describe Canada's diverse plant and animal life. 'Between ourselves,' he commiserated with his old friend Bebb, 'the *old men even in our days* seem to see better than the younger men. Note how the latter spend their time in trifling things and leave all the real work to us ... the question is who are the men of the coming generation who are to take it up ... men calling themselves botanists know much less about plants than they think they do.'[77] Macoun was thus determined to spend every field season on the ground. He saw himself as one of the few remaining members of a generation of all-round naturalists and, if at all possible, wanted to see his work through to its logical completion. 'Do you think if I stayed at home,' he asked one of his ornithological friends, 'we will get the work done as it will be done if I am there?'[78]

The Professor's continued field activities, combined with Jim's absence, meant that any special natural history reports were delayed. Macoun did manage to finish part six of his *Catalogue of Canadian Plants* but had given up any immediate hope of publishing a flora of Canada. Instead he supported Dr Britton's plan to prepare an illustrated guide of the flora of North America and offered to share his knowledge on the range and habitat of Canadian plants. He was even willing to take Britton to the field with him. His only concern was that the word, 'Canada' appear on the title page of the proposed manual. As he told Britton, he regarded his botanical work as 'getting together a grand collection of Canadian plants' as 'a foundation for others to work upon.'[79] The bird catalogue also remained incomplete. Part of the problem was that Selwyn, still bothered by Chamberlain's criticism, refused to let Macoun prepare the book with McIlwraith – it had to be a Survey endeavour.[80] But every time Macoun and Spreadborough went to the field, the Professor returned feeling that another season of collecting and observation was needed. The other major stumbling block was his work-load. 'I am at every *branch* of Natural History,' he explained to one of his many American contacts,' and I might say every part of every branch so you see my head and hands are full to overflowing.'[81] Macoun simply could not find enough time to examine Spreadborough's field notes and specimens, as well as copy information from all the published sources on Canadian birds. So although he made several attempts to complete the bird catalogue and expected to 'make short work of it,'[82] he only dabbled at it in fits and starts. Ironically, by the time he was able to get it in a near-finished state, it could not be published because of the lack of funds.

Relations between Whiteaves and Macoun, in the meantime, continued to be rather cool. This situation was largely the Professor's own doing. Collecting aquatic invertebrates whenever possible, Macoun and Spreadborough had used a small hand dredge to make a particularly large collection on Lake Erie in 1892. These specimens should have been examined by Whiteaves, who had considerable expertise in marine invertebrates and was planning a publication in the area. Macoun, however, had 'given up all hope of him doing anything' and sent part of his shell collection to G.W. Taylor of Victoria, British Columbia, an amateur conchologist who also hoped to publish on the topic.[83] Relations between the two men consequently became so strained that Macoun had a separate cabinet built for his marine collections and told Dr Selwyn that Whiteaves was not to touch them.

The Professor's relations were no better with the Survey's taxidermist, Samuel Herring. At odds with Herring ever since he joined the staff, Macoun found him to be irritatingly slow and was constantly trying to tell him how to do his job. Their relations took a turn for the worse in 1890 when Macoun learned that a box of bird skins in Herring's care had been found, destroyed by insects, in the basement.[84] From that point forward, Macoun was continually looking for a replacement. 'Everything remains here as before,' he told one prospective candidate. 'Herring has been telling some of the men that he will leave in the spring but he does not mean it. He is too well off. Should he give me a show he will go in a hurry.'[85] This tense situation certainly did not help in getting the field material mounted – it took more than three years to prepare the collection for the Banff museum. In fact, not only were Macoun's collections sent south for determination, but much of the Survey's taxidermy work was done by Ward's Natural History Establishment in Rochester, New York.

As if these various difficulties were not enough, Spreadborough and Jim Macoun laboured under the weight of personal problems. While travelling in the back of a wagon at Indian Head in May 1892, Spreadborough killed the driver when the shotgun that he used to shoot birds accidentally discharged. The Professor was 'terribly cut up' on receipt of the news, 'as it was the last thing I would have expected to have happened by you. It is quite evident it was a pure accident but a most unexpected and fearful one. As you say had it been your dearest friend you could not have helped it, so do not worry over it as you cannot mend matters now.'[86] The incident left a profound impression on Spreadborough and he never really got over it; for many years thereafter, he looked after the driver's widow and eventually married her.[87]

Macoun's son was also troubled by problems of a personal nature. His wife and his two young daughters, Helen and Mary, were often sick, and he faced large doctors' bills. Since moving to Ottawa, Jim had also become closely attached to a small circle of civil service friends that included poets Archibald Lampman and Duncan Campbell Scott. (Following the 1902 death of his first wife, Mary MacLennan, Jim married Scott's sister, Helen.) Through regular meetings with this group, Jim was introduced to British socialist thought and became an avid reader on the topic, as well as one of the leading figures of the local labour temple. He also came to deplore the sleaziness of Ottawa politics and was a persistent advocate of a patronage-free civil service. This latter crusade was sparked by his general unhappiness at the Survey. It was not that he did not enjoy his natural history duties but he had become increasingly disenchanted with his temporary status and his seemingly bleak prospects and was known to drink heavily at times.[88]

Jim Macoun's worries about his career seemed about to be remedied in 1893, as a result of his work on the fur seal controversy, particularly during the Paris arbitration hearings. So pleased were Canadian officials with his efforts that, before Jim's return to Ottawa, C.H. Tupper, the minister of marine and fisheries, spoke to him of the possibility of a permanent position in his department. 'This is an opening,' Jim wrote to his father on 17 July 1893, 'that I cannot let slip. I am now in a position I may never be in again ... I can say what I want and if I don't get it I can leave.'[89] What Jim proposed to do, since he was already handling most of the work, was demand to be appointed Survey botanist. He believed that if the appointment was not made while he had the benefit of his father's assistance and knowledge, it would never be made and 'that is a risk that I cannot affort to run.'[90] In a spirit of regret, then, he advised his father that he was determined to leave the Survey unless his situation improved.

Yet when Jim wrote to Dr Selwyn three days later, he told him of Tupper's offer but only asked to continue to be paid at the same rate he had received while on seal business.[91] He made no mention of his desire to be appointed botanist, having decided to wait until Tupper's return to Canada before he pursued the matter. It was a wise decision, for the position fell through and he was forced to remain at the Survey as his father's assistant. This turn of events must have been particularly discouraging, for despite his good work on the seal business, his position remained temporary. Jim, for his part, blamed the outcome on his political leanings; the government could hardly be expected to do otherwise, he told a close friend a year later, since he was 'a rabid socialist ... so far gone I contribute a column or two of stuff to a Labour paper here every week.'[92]

These problems of field and office did not deter the Professor from carrying out his duties as Survey naturalist. He had by now learned to make the best of any situation and was not about to be sidetracked from the task at hand. 'I certainly have had good chances,' he wrote to a British colleague, 'but if I have had I either made the chance or *worked* it for all it was worth.'[93] He probably never really understood his son's sense of frustration, and he expected Spreadborough to soldier on, whatever the circumstances. The Professor's one overriding concern was the complete listing of Canada's natural life and as far as he was concerned, this goal should be theirs as well, though clearly the task did not consume them to the extent that it did the senior Macoun. Fully prepared to keep at it until his death, he regarded his natural history work as a great service for Canada and he never hesitated to tell anyone so.

In this endeavour to put Canadian natural science on a sounder footing, however, Macoun's ties with American naturalists steadily increased. He visited the United States each winter whenever finances permitted and even had a chance meeting with President Benjamin Harrison on a trip to Washington in December 1890.[94] This continental connection was further strengthened by son Jim, who came to know a number of American scientists during his visits to Washington on seal business. Such consultation was of great benefit to the Macouns in that they kept abreast of ongoing developments. It also meant that their wide and varied collections received the attention they richly deserved. There were times, however, when this collaboration was a little one-sided. Though they sent parcels of Canadian plants to all the major American herbaria in an attempt to clear up the confusion surrounding northern and southern forms, as well as to ensure that they had a complete set of Canadian plants, Jim often complained that Ottawa received 'no decent return from any of them.'[95] American scientists also had to attend to the natural history of their own country and there were often delays in working over the Survey's material – the same problem Macoun had encountered with British specialists. He waited three years, for example, before the first lot of reptiles that he sent to Stejneger were named. Jim, meanwhile, would have to hold back the distribution of duplicate specimens until the more difficult ones had been properly identified and named. He waited patiently, however, feeling somewhat 'ashamed' in the first place to ask leading American botanists 'to give more ... time to the determination of our specimens.'[96]

The Macouns also experienced some difficulty in getting a species new to science acknowledged. Sometimes a new species sent to the United States for identification would be placed provisionally with a known species until the same new species had been found in America; the original Canadian discovery, perhaps made years before, would then be disregarded. 'All that either my father or myself want at any time,' Jim told one of their more appreciative American correspondents, 'is that when an order or genus has been revised, or our specimens used, Canada should at least have the credit of possessing the species under consideration.'[97] Even when a new species or variety was recognized from any of their collections, the specimen on which this determination was based ('type' specimen) tended to be deposited with the institution that the particular specialist was associated with and not in the Dominion herbarium. This practice was actually encouraged by Macoun. 'Keep the type specimens as long as they are of use to you as I am too busy to work critically at anything,' he told his good friend Bebb. 'I have too many irons in the fire that I cannot do half what I hope to do.'[98]

Macoun's willingness to have type specimens deposited in foreign herbaria, together with his obsession to be credited with the discovery of species new to science, got him in trouble within the scientific community in the early 1890s. In 1887, the Professor began to send his moss collections to Dr Nils Kindberg of the State College, Linkoping, Sweden. Before that date, his mosses had always been examined by the premier American bryologists: William Starling Sullivant, Coe F. Austin, and Thomas P. James. Macoun, however, was never quite satisfied with their determinations, because they tended to assume that European and North American forms were identical and that radically different specimens were simply hybrids of established species. This 'lumping,' as it was known, greatly irritated Macoun, who believed that it downplayed the significance of his field work and, equally important, his standing within the profession. He consequently used the opportunity presented by the deaths of these men to have his mosses examined by Kindberg – a botanist whose enthusiasm for species-making matched his own for finding new ones.

In working more than Macoun's moss collections for 1885 and 1889,

Kindberg described over thirty species new to science, including a new genus. The Swedish bryologist not only determined that the Professor's specimens were distinct from European forms, but that his eastern specimens differed from those collected west of the Rocky Mountains. Macoun was naturally elated. He wrote to Kindberg, 'I am astonished at my success in unearthing so many things or to put it plainer that you have been able to separate so many fine things from the mass of unworked material at your disposal.'[99] Convinced that the confusion surrounding North American mosses was finally being cleared up and that he would secure the recognition that he deserved, Macoun sent Kindberg for re-examination all the mounted specimens in the herbarium that had previously been determined by American authorities. He also began to devote almost all his botanical energies to mosses, sending at least two or three packages every week from the field directly to Sweden. By 3 September 1890, Kindberg, in consultation with Dr Carl Müller of Halle, Germany, had credited Macoun with 171 species of mosses new to science. Müller went so far as to suggest that the Canadian's work constituted an epoch in American bryology.[100]

Caught up in his own apparent success, Macoun bragged about Kindberg's determinations to a number of leading American botanists. C.R. Barnes, editor of the *Botanical Gazette*, was advised that 'much work has yet to be done on both sides of the line 49,'[101] while G.N. Best was lectured, 'with my 30 years knowledge and experience I laugh *silently* at anyone whose span of life had not begun when I was a master ... setting up for a profound knowledge of the whole subject.'[102] Largely on the strength of Kindberg's critical work, Macoun also began to sell sets of Canadian musci to interested botanists or offer samples for exchange.[103]

All this species-making or 'splitting' was soon questioned in the United States. In particular, Mrs Elizabeth G. Britton, who was associated with the New York Botanical Garden and had met Macoun in Toronto in 1889, complained that many of the supposedly new species and varieties of mosses were based on slight differences and poorly characterized. In response, the Professor freely admitted that Kindberg and his colleagues were probably guilty of making too many species, but added, 'I do put faith in their knowledge of European forms. If ours do not agree with the European ones then ours are *not* identical and are either new varieties or *species* ... I am neither a *lumper* or a *splitter* but believe that every permanent form should have a name.'[104] Despite the warnings of specialists like Britton, then, Macoun believed that he was doing the right thing – both for science and himself – and continued to forward his mosses to Sweden. He did, however, allow Mrs Britton the privilege of re-examining any of Kindberg's questionable specimens, but only as a basic courtesy. As Macoun commented on the affair, 'I am becoming very fast a real German philosopher. There is nothing real and all our old ideas must go to the wind. Where is it all going to end.'[105]

When part six of Macoun's *Catalogue of Canadian Plants* was issued in 1892, it listed 953 moss species, including 237 new to science – considerably more species than were given in the 1884 *Manual of North American Mosses* by Charles Leo Lesquereux and Thomas P. James. Fellow British geographical botanist J.E. Bagnall showered Macoun with praise, lauding the moss catalogue as one of the 'most marvellous records of industrious research.'[106] Other botanists were not as enthusiastic. In the first place, Kindberg was greatly disturbed that his name had been excluded from the title page. All copies still in the Survey's possession were consequently sent back to the printer and the cover changed. Other criticism could not be accommodated so easily. Edwin Latnall of Wilmington, Delaware, reasoned with Macoun, '... everybody gives you great and lasting credit. Everyone knows what a vast work *you* have done. But the fault lies back of you entirely, altogether with your collaborators, who certainly in a few cases have, by naming a plant differently at different times, shown, not ignorance, but carelessness.'[107] C.R. Barnes, on the other hand, was mercilessly blunt in his published review of the Catalogue: 'It is unfortunate that Mr. Macoun was not more cautious in the choice of bryologists to work up his rich collections ... The definitions of the alleged new species ... are inexcusably bad ... Altogether we must conclude that what might have been a work of the greatest value has its good distributed through a heap of rubbish which somebody must sort over before the good can be separated from the bad.'[108]

Although he consoled Kindberg that Barnes was 'a jealous young man ... scarcely worth troubling yourself about,'[109] Macoun was deeply distressed by this criticism. In a bitter outburst to Mrs Britton, he exclaimed, 'Barnes will be a failure every time. He ran before he was old enough to creep.'[110] He realized, however, that the criticism of his moss catalogue had to be answered. He had to find some expert in North America to verify Kindberg's work and thereby undo the damage that had been done to his reputation.

Mrs Britton was the obvious choice, but the two were no longer on the best of terms. She had been quite upset that Macoun had failed to include some of her notes and observations in the moss catalogue – so upset, the Professor speculated, that it was she who had set Barnes on him. Yet there was no one else in the United States that Macoun trusted with the matter and so he wrote a series of flattering letters to Mrs Britton applauding her abilities. Mrs Britton agreed to undertake the task and eventually became so severe in her criticism of Kindberg's work that the Professor offered to send his moss specimens to her first for determination.[111] Handling Macoun's voluminous collections would have jeopardized her other labours, however, and she simply continued to act as a check on what she described as Kindberg's worthless species. This was no easy chore, for Macoun did not have the type specimens – they were in Kindberg's hands. He was also in the awkward position of trying to match up his Ottawa specimens with what he thought were Kindberg's new species.[112] In the end, it took North American botanists several years to sort out the mess and 'the almost inevitable reductions that followed were borne with apparent philosophical composure, with the faith that science could be depended upon ultimately to straighten out all difficulties.'[113]

Besides relying on American naturalists for working up his collections, the professor was also being increasingly directed by them in his field work. In fact, Macoun personally requested such assistance. After outlining his proposed field work for the 1894 field season, he told Captain Charles Bendire, oologist with the Smithsonian, to 'pick out some other locality where collecting could be done to great advantage?'[114] In the process of being told how and where to collect, however, it appeared that Macoun actually took his instructions from Washington, not Ottawa. This was particularly true of his relationship with Merriam. In 1892, for example, Merriam gave his blessing to the field work that Macoun planned for that season: 'I am glad to know that you have a man at work in the northern prairie region,' he told Macoun; 'specimens of all sorts from the northern limits of the prairie country would be of much interest.'[115] It eventually reached the point where the

Professor was given an annual shopping list before departing for the field. Other American naturalists were just as demanding and, in the process, Macoun ended up providing the field data upon which they based their theoretical work. It was with a very apt comment that one of his American contacts acknowledged, 'I fear you will think I have imbibed the principles of the Monroe Doctrine, before I finish.'[116]

Macoun's relationship with American naturalists did, nonetheless, have other benefits besides the exchange of natural history specimens and related information. It proved to be rather effective in manipulating Dr Selwyn. In the winter of 1893–4, the Survey's financial picture was bleaker than usual because of the costs it had incurred in participating in the Chicago World's Fair.[117] All printing had stopped, and it appeared doubtful whether Macoun would be able to carry out his planned field work in southern Assiniboia, an area in which Merriam was also interested because of his current work in neighbouring Montana. Meeting in Boston in November 1893, the pair took it upon themselves to remedy the situation; it was evidently not the first time they had resorted to such tactics.

Their plot was set in motion in early January 1894. In a confidential letter to Merriam, Macoun advised:

Your letter to Dr. Selwyn came in due course and *as usual* he did not 'catch on' to your meaning ... He showed your letter to me but thought we knew enough about the animals of the prairie. I told him we had general knowledge but not particular ... I am not aware whether he has answered you or not and would be glad if you would write me a note regarding *your* intended work and what you would be glad *we* could do for you. Do not refer at all to this letter or say anything about our conversation but ask me or tell me anything you choose. I will take your letter to Dr. Selwyn at once and if he does not agree with me I will go to the minister and have it done ... I have fully made up my mind what to do next year or rather season, and will see that I am allowed to do the work as I wish.[118]

In response, Merriam sent a formal letter to the Professor, in which he briefly outlined his work and then asked a series of very specific questions about the limits of particular species. 'Perhaps no biological work in Canada,' the mammalogist carefully argued, 'is more important at present than determining the northern boundaries of distribution of a number of species that inhabit the Great Plains ... this work should be done at the earliest possible moment.'[119] Selwyn took the bait. On 8 February, Macoun reported to Merriam that Selwyn would not stand in his way if a detailed natural history examination of the prairies was thought necessary.[120]

Before Macoun left for the west, there was almost a falling out between the two conspirators. In December 1893, Macoun received an enthusiastic letter from Samuel Rhoads of the Academy of Natural Sciences in Philadelphia, offering to prepare a manuscript on the mammals of British Columbia based on the professor's field collections.[121] Macoun would probably have ignored the letter if he had not received at the same time a note from Merriam in which he advised that the determination of specimens would require more time because of his increased work-load.[122] In the new year Rhoads continued to write to Macoun, emphasizing the need to press ahead with the project and warning that Merriam would never return his mammal skins.

It was obvious from Rhoads's letters that he was trying to divert Macoun's specimens away from Merriam. Constantly seeking recognition of his field efforts, however, the Professor decided to test the validity of Rhoads' accusation. Citing the need to examine his mammal specimens before he took to the field again, Macoun asked Merriam on 29 March 1894 to return all his skins within two weeks.[123] Five days later, before receiving a reply, he dashed a quick note off to Rhoads, suggesting that he would send him some of the specimens as soon as they arrived back in Ottawa.[124] Merriam's reply arrived the following day. In a brief, tersely worded letter he hinted that he was aware of Rhoads's behind-the-scene intrigues and that he no longer wanted to examine any more Survey specimens.[125] The Professor was thoroughly shaken by the mammalogist's note; he had become extremely dependent upon Merriam's assistance. He did his best, nonetheless, to put up an innocent front, arguing in a return letter that 'there is no truth to the statements you have heard.'[126] Merriam did not take issue with the explanation but seemed satisfied that he had made his point. When he returned the specimens he tried to repair the damage to the friendship by suggesting that he might prepare a special report on Canadian mammals.[127] Macoun was greatly relieved and promised to return the mammal skins in the fall. The incident had such an impact on Macoun that two years later he still felt it necessary to assure Merriam that he was the first mammalogist to see his collections.[128]

The 1894 field season was to be the first of three summers that Macoun planned to devote to the natural history of the southern prairie region. From the beginning, however, the trip was beset with financial problems. Because Survey funds were nearly exhausted for the fiscal year, Macoun was not permitted to leave for Medicine Hat until late May. Here, he joined Spreadborough who had been there at his own expense for the preceding six weeks, making notes and collecting specimens. For the next month, the two men immersed themselves in the natural history along the CPR main line between Medicine Hat and Swift Current, concentrating their efforts around Crane Lake. In all his years collecting throughout Canada, Macoun had never before seen such a rich bird population as he found at this prairie lake north of the small hamlet of Piapot. It was an ornithologist's heaven, so much so that the pair suspended all other collecting activities and, wading for several days along the shores of the lake, retrieved over 1,000 eggs from the nesting waterfowl. These activities came to an abrupt end on 2 July 1894, when Macoun was ordered to return to Ottawa immediately. Dr Selwyn's attempt to have additional funds voted for the Survey had been unsuccessful, and he was warned in mid-July by Thomas Daly, the new minister of the interior, that he could not run a deficit at the close of the financial year. With a debt of $20,000, he had no choice but to cancel all field operations.[129]

Macoun returned to Survey headquarters to find that his son had the herbarium in better shape than it had ever been. Just one year earlier, when Jim finally reported back to the Survey, the botanical work that he had left behind was still waiting for him, along with that which had accumulated during his two-year absence on fur seal business. He himself added to this backlog; never one to miss the opportunity to investigate new areas, he had collected plants on the Pribilof Islands, drying his specimens in the engine room of his steamer.[130] Jim consequently spent the rest of the summer of 1893 trying to clear up all the outstanding botanical work, but by early November there was still material from 1891 to be processed. At first, the professor assumed that Jim could simply work overtime; after further thought, however, he finally realized that his son's time was far too valuable to be spent on such routine chores as labelling, mounting, and placing plants in the herbarium. He therefore asked Selwyn for temporary clerical help – something that had been needed since the expansion of Macoun's duties – on the grounds that he had been without assistance for the past two years.[131] Given the Survey's poor financial shape, such a request was unlikely to be fulfilled. Macoun, however, was prepared 'to work the political machine for all that is worth,'[132] and he had a close friend, Ottawa MP, Sir James Grant, speak to the minister of the interior on his behalf. In January 1894, the botany division secured the services of a female clerk for one day per week. With this clerical assistance, Jim made such rapid progress that by May 1894 he decided to re-examine all the Canadian plants in the herbarium.

The other natural history work of the Survey did not fare so well. It was in such arrears that the faunal specimens from the previous field season were still unpacked by the time Macoun took to the field again.[133] Selwyn's decision to recall the Professor from the field thus actually benefited the natural history work of the Survey. The restrictions on travel left him with little alternative but to sit down and sort and arrange his various collections to find out exactly what he had. This previously unachieved degree of organization was accompanied by a new name for the Survey's botany division; as of late 1894, it became known as the natural history branch. This name change was a belated recognition of the broad natural history work Macoun had performed under Survey auspices, particularly since he had been promoted to Survey naturalist in 1887. Bent on uncovering every distinct form of Canadian natural life, he had roamed through forests, across prairies, and over mountains from ocean to ocean. In this quest, Macoun saw himself as 'a clear sighted collector.'[134] He was no ordinary field naturalist though. Convinced that he was 'going to collect new species ... no matter where I go in the country,'[135] he insisted on making large flora and fauna collections. He was also willing to take whatever steps were necessary to see that nothing interfered with his field operations and demanded the same kind of devotion from his co-workers, his son and Spreadborough.

This emphasis on field work and the search for new forms had its costs. At a time when natural science was becoming increasingly specialized, the natural history work performed by the Geological Survey of Canada continued to be concerned with inventory and description. In the performance of these duties, moreover, the Survey had become increasingly reliant upon American assistance – not only for the determination of specimens but for the direction and scope of field work. Indeed, the dependence on Europe that had characterized early North American natural history study had simply been replaced, in the case of the Survey, by a new dependence on the United States.

Macoun saw nothing wrong with the direction that his work was taking. Since he had neither the expertise nor the time to work up his large, varied collections, he was quite happy to have them determined by some of the leading authorities in the United States and, to a lesser extent, Europe. That Kindberg named so many species or that Merriam and Rhoads fought over the right to examine his specimens was regarded by Macoun as confirmation of the good work he was doing in the field. He was also opposed to the direction that natural science was moving in and fully expected future Survey natural history work to be characterized by the same general wide-ranging field work that was the hallmark of his own studies. 'We are breaking the ground,' he told one of his many correspondents, 'and I look forward to the day when real enthusiastic workers take to the field and do honest work.'[136] In fact, the major problem with the Survey's natural history endeavours from Macoun's perspective was that the value of his collections could not be appreciated because there was no place to display them at Survey headquarters. Jim was constantly stashing material away 'in every hole and corner of the building,'[137] and it sometimes took an entire day to find something. The need for a new building with proper museum facilities was consequently one of the foremost concerns for Macoun.

Towards a National Museum

John Macoun's scramble to uncover the natural life of half a continent, together with the field activity of other Geological Survey officers, had an unanticipated impact on institutional development in Ottawa. Each field season, these men not only observed and recorded but also returned with any interesting or unusual geological, ethnological, and natural history specimens. Although this collecting activity was not officially required until 1877, the Survey soon became the custodian of what was generally acknowledged as a national collection – a kind of 'national attic where things too good to throw away are kept safe.'[1] By the mid-1890s, however, in the course of exploring Canada's frontiers, the agency had outgrown its facilities; it could not absorb any new material, let alone safeguard the wealth of specimens it had accumulated over the past half century.

This situation was especially disturbing to Macoun. He strongly endorsed the idea of a national museum as a practical, as well as an educational, institution in which Canada's great resource heritage could be advertised. Yet he now faced the frustrating situation at the Survey where the want of adequate museum space was, in his own words, 'a greater hindrance to progress than the want of assistance.'[2] Securing a new home for the Survey and its museum, however, would be a difficult task. With the victory of the Wilfrid Laurier Liberals in July 1896, the Survey had come under the responsibility of a dynamic minister of the interior, Clifford Sifton, who was more immediately concerned with immigration and western settlement policies than with the welfare of the Survey. In fact, far from being worried about the

Survey's facilities, Sifton was quite content to leave the department to make do with its limited budget for the time being.[3] It consequently fell to the Survey's new director, Dr George Mercer Dawson, and other interested individuals and groups to convince the Liberal government of the absolute need for a national museum. In this campaign, they wisely chose not to play politics with the stubborn Sifton and opted for the more moderate course of demonstrating the inadequacy of the Survey headquarters on the one hand and the practical economic value of its collections on the other. There was much at stake. As far as Macoun was concerned, the natural history branch would never amount to anything without appropriate work-room and display facilities.

Macoun's strong support for a national museum was a logical outgrowth of the kind of natural science he practised. Although he emphasized field work to the exclusion of detailed study in the laboratory, the act of collecting, classifying, and listing was not an end in itself. The results of his annual outings had to be mounted, labelled, and exhibited in some suitable fashion in order to reveal the great natural wealth that God had placed at Canada's disposal; otherwise, the significance of his field collections, no matter how unique, would not be fully realized or appreciated. Museums and other similar depositories were therefore regarded by Macoun as important indicators of the kind of inventory work that was being done in the field.

The critical importance the Professor attached to museums was evidenced by the large herbarium that he had maintained almost from the beginning of his botanical studies. It can also be seen in his preparation of plant displays for international exhibitions in the 1870s, as well as his participation in the 1886 Colonial and Indian Exhibition. The best example of the link that he drew between field work and the preservation and display of specimens, however, was his involvement in the Ottawa Field-Naturalists' Club. Formed in March 1879 to promote the systematic study of the natural life of the Ottawa region, the club was but one of several such organizations in Europe and North America at the time. A corresponding member of the club from its inception, Macoun regularly took advantage of his visits to Ottawa to attend club meetings. Here he met men and women with attitudes to natural history that matched his own. Among them was James

Fletcher, an enthusiastic amateur naturalist who worked in the Library of Parliament and was later named Canada's first Dominion entomologist. Fletcher was forever preaching the joys of nature study and the wholesome effect it had upon the investigator. At the club's inaugural address, on 24 November 1879, he captivated his audience with a stirring portrayal of a naturalist: 'No one looks upon the world so kindly as he does; no one gives so much attention to, or takes so much enjoyment from, the country as he does, and he holds a more vital relation to nature, because he is freer, and his mind is more at leisure.'[4] Macoun made a similar claim while attending one of the club's winter soirées in March 1880. Introduced to the meeting at the conclusion of that evening's presentation, he 'made a vigorous speech, setting forth the claims of Natural Science as a means of training both mind and body to greater and more prolonged activity than can be attained by any other course of study.'[5]

Upon moving to Ottawa, Macoun immersed himself in club activities. Apart from holding various administrative posts, he was a frequent speaker on the annual winter lecture program, usually recounting his latest adventures in the field, complete with mounted specimens and lantern slides. He also served for varying lengths of time as one of the group leaders of the botany, entomology, and ornithology branches of the club. These small early evening gatherings, originally organized in 1881 to encourage in-depth study and discussion in a particular field, quickly evolved into a regular feature of the Macoun household at 98 James Street, where one could always engage in lively, spirited conversation. He also conducted an afternoon series of natural history talks at the Ottawa Normal School or YMCA in which he emphasized the best methods of hunting, collecting and preserving specimens. 'I give popular facts,' he told fellow botanist Dr Thomas Burgess of London, Ontario, 'and do not burden my hearers with details which are generally dry but I am careful to give scientific facts and not twaddle.'[6] Often he would meet his young charges at the Survey building on Saturday afternoons to examine the museum's holdings; here he used the occasion to argue that a species list was worthless without the specimens and that any field collection would be of little value unless it was properly maintained and made available for study and comparison.[7]

It was ironic that Macoun made his comments about the importance of a museum collection in the same building where the storage and display facilities for his own natural history work were extremely poor. In fact, the Survey museum had been a highly controversial topic from the time Macoun joined the Survey as Dominion botanist. The 1877 bill that had made the agency a branch of the Department of the Interior and greatly enlarged its duties had also called for the removal of the Survey and its museum from Montreal to Ottawa at some future date. This provision, more than any other feature of the new Survey Act, had sparked considerable debate during second reading, as representatives from the two cities lined up against one another. Montreal MP's could understand the removal of the Survey's offices to the nation's capital; but the loss of the museum, which represented nearly forty years of Survey collecting efforts, was bitterly opposed. Bernard Devlin of Montreal Centre confessed that 'he was at a loss to know what crime Montreal had committed.'[8] Ottawa representatives were equally determined to secure the Survey museum. Sir James Grant, a former Conservative MP and chief physician to the governor-general, had privately advised Prime Minister Macdonald shortly after his return to office that, 'The Montreal building ... is quite inadequate for the needs of the country ... and the change to Ottawa would be exceedingly gratifying to members of both Houses.'[9]

Despite the contentious nature of the proposed transfer, the new Macdonald government decided to go ahead, and in the early spring of 1881 it purchased the former Clarendon Hotel in the old town market area just east of the Parliament buildings. Montrealers, however, refused to admit defeat. On 16 March 1881, a deputation of mostly Conservative senators and MPs belonging to the Montreal Board of Trade called on the prime minister to delay the dismantling of the museum in order that some mutually acceptable solution could be reached. Sir John 'promised that immediate attention should be given to the matter,'[10] but the move to Ottawa proceeded. Montrealers were left to console themselves with a grant from the Redpath family to expand the McGill University museum.

Even after the Geological Survey had settled into its new Ottawa home at Sussex and George Streets, the issue was still very much alive, although it had now evolved into a party question. When, in May 1882, Matthew Gault, the Conservative MP for Montreal West, remarked that the new Survey headquarters were 'most admirably situated' and that the museum was receiving more visitors than it had in Montreal, Timothy Anglin, the former Liberal speaker of the Commons, shot back: 'The number of visitors does not prove the place is a proper one, and it does seem an extraordinary thing to spend $5,000 in heating up a museum in which there are nothing but mummies and fossils. The building looks out on what? On a cart stand, a pile of filth, with a stench pervading the atmosphere all the time, with dust in the summer and mud at other seasons.'[11] Alexander Mackenzie was more subtle though no less biting in his criticism, when he caustically remarked, 'still it must be admitted that it is not bad for a fossil government.'[12]

Party politics aside, the criticism levelled at the new Survey headquarters was not without foundation. The lower town location was not the ideal site for what soon would be regarded as Canada's 'national collection,' and it would appear that the government had probably purchased the Clarendon as a favour to its owner, Conservative senator James Stead, who had been forced to close the luxury hotel because of lack of business.[13] As for the building itself, it had played a prominent role in Ottawa's cultural past. It was the home of the city's first public reading room and the headquarters of the Bytown Mechanics Institute. In March 1880, it had also hosted the inaugural exhibit of the Royal Canadian Academy of the Arts – the forerunner of the National Gallery. The Clarendon and its predecessors on the site, however, had been designed for accommodation purposes; even during the Fenian raids in the late 1860s, the structure had served as a temporary barracks.[14] It was consequently not long after its occupation by the Geological Survey that its shortcomings were realized.

In his first official report following the move to Ottawa, Dr Selwyn complained that there were already more specimens than available museum space. 'If the recent natural history branch of the Survey is to be carried out,' he warned in reference to Macoun's appointment, 'additional accommodation for work-rooms and exhibit space is now required.'[15] The space problem, in fact, was so bad that with only one small room set aside for natural history, the Professor was obliged to work out of his Ottawa home during his first two years in Ottawa.[16] In responding to Selwyn's grumbling, Senator David Macpherson, the new minister of the interior, recommended to the prime minister in April 1883 that an additional $10,000 appropriation to extend the museum galleries would be the 'most convenient as well as the cheapest' way to resolve the problem.[17] No such sum, however, found its way into the supplementary estimates for 1883. The depressed state of the Canadian economy was undoubtedly a factor. Faced with declining revenues, the Conservative government was more eager to reduce expenditures than to assume additional burdens. There was also the widespread feeling among parliamentarians that the Survey was in need of reforms of a more fundamental nature, not just improved facilities.

Selwyn was not deterred by the government's failure to act on the matter and, throughout the 1880s, he regularly drew attention in his annual reports or public lectures to the need for more museum space for 'the proper arrangement, preservation and exhibition of the constantly augmenting collections.'[18] Nor was he alone in this campaign. In his introductory remarks to the Interior report for 1886, Deputy Minister A.M. Burgess argued that the creation of a national museum merited serious consideration. 'Excellent results have already followed from the removal of the headquarters of the Survey and the museum to Ottawa,' he remarked, 'but the full advantage of the change will not have been reaped until quarters better suited for the display of this collection, and greater facilities for office work, have been obtained.'[19]

Never having known what it was like to have adequate space for his field collections since joining the Survey, the Professor shared these concerns. In the same April 1883 letter to the Ottawa *Citizen* in which he reported that the government was not interested in purchasing his private herbarium, he lectured:

rest assured that were provisions made for accommodating a collection of natural history specimens very little time would elapse before a creditable representation would be got together. Until there is some place where such objects can be properly arranged and classified there is no incentive for anyone ... When public sentiment becomes properly aroused in regard to the value of a natural history museum and the government is authorized by the people's representatives to establish such an institution and the building in course of erection then and only then is it time to ask the staff of the Geological and Natural History Survey to obtain materials for such an establishment ... As the museum now stands such collections would have to be put away in boxes, at the risk of being destroyed by damp or insects. We live in hope and anxiously await the dawn.[20]

Macoun was also distressed by the amount of time and money that Ottawa spent on the temporary Canadian exhibit at the Colonial and Indian Exhibition and returned home determined to correct the pitiful situation in Ottawa. During his presidential address to the Ottawa Field-Naturalists' Club on 13 January 1887, he confessed, 'my English visit has been so convincing that I consider silence concerning our position a virtue no longer, and take the present occasion of laying our case before an Ottawa audience in the hope that we may arouse attention to a question that is truly a national one.' He then went on to advocate 'one grand museum for Canadian science which will be a credit to our city and a lasting monument.'[21]

Similar sentiments were expressed at the annual meetings of the Royal Society of Canada. In his 1883 presidential address, Sir William Dawson spoke of the need for a national museum to represent the resources of the Dominion, as well as to act as a scientific workshop for practical research. 'The expenditure,' Principal Dawson concluded, 'must be undertaken within a few years if Canada is to take its place worthily among the civilized nations of the world.'[22] To this end, the society drafted a circular appealing for geological, biological, and ethnological specimens to be sent to the Survey. Governor-General Lord Lorne also called upon Prime Minister Macdonald to have plans drawn up for a national museum building and asked to have the ceremonial first stone laid during the society's next meeting.[23] Such noble efforts were not completely selfless, for the Royal Society was in need of a home itself and fully expected to share any new building with the Geological Survey.[24]

Despite these appeals for a new home for the Survey and its museum, the Macdonald government continued to be more concerned with the nature of the Survey's work than the provision of more adequate facilities. In its concern for utility, the government seemingly overlooked the practical value of the museum collection in illustrating the economic resources of the country. Not one of the museum advocates, however, had directly identified their cause with the Conservative national policies of development and integration. Thus, apart from the constant repairs, the only major change that was effected by 1887 was the addition of four new offices on the top floor of the adjoining building. Within two years, the men occupying these offices petitioned Selwyn to 'take steps to relieve them of the feeling of insecurity arising from the weak and sinking conditions of the floor and roof.'[25]

The Survey museum, meanwhile, continued to receive an ever-increasing number of visitors. In its first full year of operation in Ottawa, the recorded attendance was 9,549; the previous twelve-month high in Montreal had been 1,652.[26] By 1888, this figure had almost doubled, causing Selwyn to argue in his annual report that the public interest would be best served if the museum were open on Sundays. He reasoned: 'a museum ... is essentially as much a place of instruction as are the Churches and Sunday schools ... in the museum the *work*, and in the church and school the *word*, of the Creator is expounded. This admitted there seems no obvious or intelligible reason why one establishment should be closed and the other open on the Sabbath.'[27] This attempt by Selwynto secure longer museum hours did nothing to resolve the basic problem of the inadequacy of the Survey building. It did, nonetheless, serve to demonstrate the popularity of the museum – a phenomenon that was largely a consequence of its holdings. By the late 1880s, the unpretentious building on Sussex Street was the custodian of one of the premier museum collections in Canada, ranging from dinosaur fossils to ceremonial Indian dress. When compared to about thirty other major Canadian museums that had been established by universities, local natural history societies, or various government levels since the mid-nineteenth century, the Survey museum had the largest number of specimens. Only the Redpath museum in Montreal came close to challenging this position. At the same time, the Survey collection was a poor cousin to those of its American counterparts, such as the American Museum of Natural History, the California Academy of Sciences, or the Smithsonian Institution. Many Canadian items, moreover, had found their way into large American museums. The Field Museum in Chicago, for example, hired collectors – some would say grave robbers – to secure ethnological material from along the British Columbia coast.[28] Ottawa would have to address this problem in the future. The most immediate concern for the Geological Survey, however, was that there was no place to display material already on hand in an attractive manner. Although there were now separate rooms set aside for birds and ethnological material, the museum galleries on the second and third floors were so overcrowded that many of the heavier geological specimens had to be kept in the cellar. Macoun's own situation was woeful. His first-floor office was nothing more than a 'narrow space,' so that his herbarium cases were dispersed up and down the building's corridors.[29]

Following Macoun's promotion in 1887, Ontario ornithologist Thomas McIlwraith argued that the new Survey naturalist was now 'responsible for the condition of the museum and the Country will look to have it in such shape as will not bring reproach on us ... *It should be done* and *it is you that should do it* and now is the time to set about it.'[30] The Professor, meanwhile, had become extremely discouraged about the prospects for a new Survey home and frankly admitted, 'I saw the South Kensington arrangement when I was in England but never hope *even to see a building* where such an arrangement could be carried out here.'[31] Instead, Macoun was more worried about the safety of his collections. In his annual report for 1889, he cautioned that his work-room was 'so crowded by inflammable material that a spark or the dropping of a match, would in a few minutes, cause the destruction of specimens of inestimable value, which could never be replaced.'[32] Dr Selwyn, as part of a new strategy to emphasize the practical value of the collection and the critical need of protecting it, voiced the same concern several months earlier. In an April 1889 interior memorandum, he put himself on record that the present site of the Survey museum was unsuitable because it was subject to the 'constant risk of total loss by fire.'[33]

Selwyn's new tactics received considerable outside support from the Canadian mining industry, which came to regard the Survey's collections as an excellent vehicle for promoting the Dominion's mineral wealth. In the February 1893 issue of the *Canadian Mining Review*, an editorial under the title, 'The National Museum,' noted that: 'The time has come for the Government of Canada to house these valuable collections properly, place them advantageously and in a fire-proof building in some spot where the general public and citizens can easily reach it. The money could not be better spent ... A national collection would tend to ... cement the bond of union between the provinces, ... and in an harmonious whole all would tend to the upbuilding of a bright and prosperous Dominion.'[34] That same month, the annual meeting of the General Mining Association of the Province of Quebec unanimously passed a resolution along these lines. In particular, it drew attention to Macoun's earlier observation that the federal government spent thousands of dollars on foreign exhibitions while neglecting the display of Canada's resources at home.[35]

The call for a national museum building on economic grounds was not ignored by members of Parliament. During consideration of the Survey appropriation in June 1892, there was general agreement on both sides of the House with the comment of David Mills that the Survey building was 'not convenient, not very suitable and not very safe.'[36] The following year, in an apparent response to lobbying from the mining industry,[37] the new Conservative minister of the interior, Thomas Mayne Daly, privately admitted to J.A. Ouimet, the minister of public works: 'I am very much impressed with the absolute and immediate necessity of ... a proper fire-proof building for the safe and convenient housing of mineral specimens and other valuable exhibits contained in our Geological Museum ... and I really think the Government would be derelict in their duty were they not to take this matter into their serious consideration at once.'[38] He asked Ouimet to instruct chief architect Thomas Fuller to prepare plans and then insert the cost of the building in the estimates for the following year. 'The building should be of such proportions and commanding appearance,' Daly noted, 'as to be worthy of our Great Dominion and of the high and important aims it is intended to serve ... In short the whole structure, in my opinion, should be built with a view to meet not only the present necessities for such a building but also as far as possible the demands for all time to come.'[39] Such a costly undertaking was politically impossible, however, given the fact that the Canadian economy was experiencing another of its periodic trade and financial depressions. Many government departments were starved for funds, and the largesse that had characterized Conservative public-works activities in the 1880s had been replaced by a measured parsimony.[40] When Sir James Grant, at the urging of the Ottawa Field-Naturalists' Club,[41] consequently raised the matter in the House in April 1894, he was informed by Ouimet that the administration of Sir John Thompson had no plans to proceed with a new geological museum at that time.[42]

This refusal of the Conservative government to commit itself to a new Survey building came as a great blow to Macoun. He had fully expected that the initial financing for the project would be approved and that actual construction might begin as early as the summer of

1894. He was also quite anxious about the continuing space problems, which had become markedly worse since he had expanded his collecting efforts to include all forms of biological life. 'The material I collect at present is just stowed away,' he disclosed to ex-rival Montague Chamberlain, 'awaiting the stirring of the *museum* waters.'[43]

The campaign for a new Survey building entered a new phase in early 1895 with the appointment as the new Survey director of Dr George Mercer Dawson, the son of the influential McGill University principal, Sir William Dawson. The appointment of the popular, diminutive Dawson had been anticipated several times since 1888,[44] but it was not until 7 January 1895 that he assumed the position. In late 1894, the new Conservative ministry assumed that Dr Selwyn's departure for England on a three-month leave of absence was a prelude to his retirement after twenty-five years at the helm and simply went ahead and issued his superannuation orders. When Selwyn turned up at his office on 10 January 1895 to find Dawson installed as his successor, he consequently had little choice but to step down and, despite the government's gross mishandling of the matter, did so with relative aplomb.[45] In fact, the one person who was most upset by the turn of events was geologist Robert Bell, who had put forward his own claims for the position.

Dawson's appointment was timely, in that he was a man of extremely broad scientific interests who, despite his training as a geologist, had always been keenly interested in the natural history work of the Survey and had collected much of the ethnological material on display. He also knew first hand what constituted a good museum, having made an exhaustive tour of Europe's great museums in the summer of 1882, and had been a constant advocate of the need for better facilities. Just prior to taking over from Selwyn, he had told the Royal Society at its 1894 meeting, 'This collection is not merely a matter of record ... it is fitted to become also a great educational – and I may add – a great advertising medium in regard to the mineral resources of the country.'[46]

Within a month of his promotion, Dr Dawson asked Macoun to take stock of the Survey's flora and fauna collections and distribute his bird desiderata among collectors.[47] The Professor suspected that these instructions were part of a larger move on Dawson's part to secure a new museum building.[48] He was right. On 21 March 1895, Dawson wrote to the minister of the interior about the structural weaknesses of the Survey building and the dire need for more suitable accommodation. He warned Daly, 'The building is not strong. We have already had to move all heavy specimens from the top floor.'[49] Daly, for his part, was in complete sympathy with Dawson and instructed him to ask Ouimet to make provision for a new Survey building in the estimates for the next parliamentary session so that the matter would come up for discussion in cabinet. Such a request was even more unrealistic than it had been two years earlier. Not only did the economic climate remain dismal, but the once great Conservative ship of state was on the rocks and badly listing, thanks largely to the inept leadership of Macoun's old Belleville friend Sir Mackenzie Bowell. The politically astute Dawson should have been aware that a new building was an impossibility. At the very time that he was making his plea for better accommodation, he was fighting to prevent a further reduction in the Survey's annual operations grant, which had already been cut by $7,000 the previous year.[50] Then again, he attached such great value to the museum collection that one of his first acts as director was to limit the loan of material.[51]

The impoverished state of the Survey in the spring of 1895, combined with the uncertainty as to whether there would be a general election or another session of Parliament, caused Dawson to continue his predecessor's policy of restraint. Most printing remained at a standstill, and the summer's field operations seemed likely to be severely curtailed again. Macoun wanted to continue his natural history survey of the southern prairie region but realized that he would not get to the field until mid-summer, if at all. 'I may say that in my experience there never was a time when money was so scarce,' he advised a job seeker.[52] During the recent parliamentary session, however, many members had expressed great concern over a continuing drought in southern Assiniboia and Alberta and had begun to question the wisdom of earlier decisions to settle the region. By coincidence, this region was the same country that the Professor wanted to examine. While other field parties were consequently delayed until the Survey's financial troubles were resolved, Macoun was dispatched to the prairies in early May under instructions to determine the causes of the prolonged dry spell as well as to discharge his regular natural history duties. As in past years, his field assistant,

William Spreadborough, had been sent ahead to collect around Moose Jaw.

While his father was absent in the field, Jim Macoun was forced by the shortage of funds to remain in Ottawa and attend to routine office duties. Botanical matters consumed most of his time, as he was either identifying collections that arrived almost daily from different parts of the country or getting together packages of plants for distribution to other institutions or individuals. The days were often long and tedious. 'I have never been so busy,' a tired Jim wrote to his good friend Theo Holm, 'since I have been on the staff of the Geological Survey and have to crowd into a day the work of two or three.'[53] At one point he lashed out at Holm when he and his father were criticized for not keeping up with the current botanical literature: 'Between 1882 and 1890 I spent 73 months under canvas – more than six years,' he wrote in their defence; 'during the winter we have had to arrange our specimens and get them determined some way, keep our herbarium ... get out plants for mounting, distribute specimens, attend to correspondence and determine specimens from all parts of Canada, besides incidental work of all kinds.' He continued: 'You can imagine, if you will think of this for a moment, how very little time this has left us for *book* study. On the other hand we have managed to do much original work.'[54]

Macoun returned to Ottawa that August, confident that 'the permanent drying up of the country was a myth.'[55] These results had been initially unexpected by Macoun. At the beginning of his investigations, as he and Spreadborough worked their way southwestward from Moose Jaw to Old Wives Lake, they found the effect of the drought on the land staggering. While camped along the hills of the Missouri Coteau, the Professor recorded in his notebook: 'Drought showed in the seedless grass, the cracked sod and parched soil and dried up ponds.'[56] He was quite worried that if a fire swept through the region 'in its present weakened state the pasture lands would be reduced to lasting sterility.' He reasoned, almost desperately, 'This season ought to be fairly wet. I pray God it be.'[57] The spring rains did come that year. Beginning in late May and continuing into July, there were heavy thunder showers almost every other day. The results were astounding. Except for a dry section south of the Cypress Hills, the prairie came to life before their eyes as they travelled westward along the international boundary towards the foothills. They responded by collecting as wide an assortment of the region's natural life as possible. Some days the findings were so large – for example, on 29 July Spreadborough returned to camp with seventy-one species of birds and mammals,[58] – that they had trouble putting up all the specimens.

That winter, Macoun and his son also attempted to take advantage of Dawson's long-standing interest in their work to secure Spreadborough's appointment to the new natural history branch. This idea had originally been raised by Jim in 1893 but was not acted upon because his father had already arranged for Spreadborough to spend the winter collecting birds on Vancouver Island. Since then, the camp-hand-turned-naturalist had proved himself such a capable and devoted field worker that the Macouns decided to try to arrange for him to work year-round on Survey business as a member of the temporary staff. In this way, Spreadborough could spend his winters in Ottawa working up his summer's collections instead of trying to pick up odd jobs in Muskoka to tide him over until the start of the next collecting season. In January 1896, then, both father and son wrote to Dawson, who passed their request on to the minister of the interior.[59] Nothing came of the proposal, however, for the Conservative government was in the midst of a grave internal crisis. Faced with a major cabinet revolt over his handling of the Manitoba School Question, Prime Minister Bowell had been forced to step aside for that old political warhorse Sir Charles Tupper, who assumed the unenviable chore of guiding the government's remedial bill through the final parliamentary session. The Survey soon faced a crisis of its own when the life of Parliament expired before the Survey's appropriation for the fiscal year 1896–7 could be approved. Despite the department's history of economic hardship, this state of affairs was unprecedented. It was only after receiving approval from the new minister of the interior, Hugh John Macdonald, that Dawson dispatched the summer's field parties, with the warning that they were subject to recall at any time.[60]

Macoun spent the 1896 field season on the prairies once again, collecting mostly plants. For the first time in several years he was without field assistance. Always eager to secure specimens from the more remote regions of Canada, Macoun had arranged for Spreadborough to act as naturalist to the A.P. Low expedition across the Ungava peninsula from Richmond Gulf to Fort Chimo. His son, meanwhile, had spent the first three months of 1896 in London on Behring Sea business and was then sent back to the Pribilofs in May to conduct further studies as a British agent. Jim could not have been happier. Before heading off to the sealing grounds, he had written to one of his American botanical correspondents, 'My li'l star is in the ascendant again.'[61] The Professor's summer on the prairies was equally eventful. Setting up base at Brandon, he found on 12 June 1896 an extremely rare Connecticut Warbler nest in a tamarack swamp at Sewall, south of Carberry, Manitoba. This was the second most important oological find of Macoun's career – the first being his discovery of a number of passenger pigeon nests along Manitoba's Waterhen River in 1881.[62] He also made some interesting observations on the potential of the North Saskatchewan country when he travelled northwest to Prince Albert. 'I ... am quite satisfied that 300 miles north of the boundary the climate is as good if not better,' he reported. 'My three seasons' experience have convinced me that while the prairie is even richer and more valuable than we believed it to be, the brush and aspen district to the north of it is better suited for immediate settlement.'[63] These comments were not altogether surprising given his previous field season's experience in southern Assiniboia. It was also his first detailed study of the land around the forks of the Saskatchewan since he had passed quickly through the region as a member of the 1872 Fleming expedition. Naturally, he lauded it. Canada in Macoun's eyes had great potential, and each region would receive its due recognition once it had been examined by him.

When the Professor returned to Ottawa he found many of his old Conservative friends warming the Opposition benches, displaced by the Laurier Liberals who had assumed office on 11 July 1896. For the sixty-five-year-old Macoun, it was a logical time to contemplate retirement or at least a reduced work-load. The timing was all wrong though. The final and seventh part of the *Catalogue of Canadian Plants* and the long-delayed bird catalogue were still a few years away from publication. There was also the question of who could assume the work. Macoun continued to believe that young naturalists were not doing enough work in the field and that Canadian natural history would be best served if he continued to do the work himself. In a vain letter to Mrs Britton on the state of natural history work in Ottawa he complained, 'Our Club [OFNC] in this city is all wind. My son and myself are the only workers here and we get no help from outside. I have tried every means to stir them up but have not succeeded yet.'[64] Then there was the matter of a national museum building. It made little sense to Macoun that, after having his wide and varied collections attended to by some of the world's top specialists, he should then, for lack of adequate space, have to hide them deep within the bowels of the Sussex Street building; as illustrations of the country's unlimited potential, his collections deserved to be prominently displayed in a building in keeping with their value. Thus, despite the change in government, Macoun had no intention of slowing down, but was anxious to see whether the Laurier Liberals would be any more willing to undertake construction of a national museum. In the meantime, never one to let the chance to get away to the field slip by, he explored along the route of the recently completed Parry Sound Railway until the weather made collecting impossible.

After the transition in government had taken place, Dr Dawson diligently set to work to win over the new administration to the cause of better Survey facilities; it would not be an easy job, for the Liberals upon assuming office took an even more stringent position on public buildings than their Conservative predecessors.[65] On 9 September 1896, in replying to a memorandum on the proposed establishment of a Dominion Bureau of Mines in Montreal, Dawson used the occasion to outline to the secretary of state and acting minister of the interior, R.W. Scott, the problems of the Survey, in particular the inadequate housing and exhibition of its 'National Collection.'[66] In his annual report later that year, he noted that the 31,595 visitors to the museum that year confirmed the need for a modern, more spacious building. That the need was urgent was underlined by the occurrence of a fire next to the Survey building in the past summer. Dawson cautioned: 'The collections, embracing as they do more than 2,000 unique "type" specimens, with the entire supply of reports and maps, and the manuscripts and notes representing over fifty years of work, would constitute an irremediable loss to the country if destroyed.'[67] Dawson saved his best ammunition, however, for his new chief, Clifford Sifton, a Brandon lawyer and former Manitoba attorney-general, as well as the most capable man to hold the Interior portfolio to date. In a letter dated 26 January 1897, Dawson recounted to Sifton that the Department of Public Works, through Thomas Daly's earlier efforts as minister of the interior, had been made aware of the absolute necessity of a new Survey building and was busy preparing some preliminary drawings. He also explained that he had been in touch with various renowned museums throughout the world over the past few months about the requirements for a national museum building in Ottawa. He concluded by suggesting that the minister contact his counterpart at Public Works and have building specifications worked up so that the cost and other associated aspects of the project could be considered further.[68]

Dawson did not stop here. Through his efforts, the matter was raised on the floor of the House of Commons on 17 May 1897, when N.A. Belcourt, the Liberal MP for Ottawa, introduced a motion calling for copies of all petitions and letters that the government had received in support of the erection of a national museum. In justifying this motion, Belcourt first drew attention to the international reputation of the Survey museum, quoting extensively from letters from foreign scientists that had been supplied by Dawson. He then went on to inform his fellow parliamentarians that 'a very large portion of the collection is now stored away for want of space, some specimens in the attic –some in the cellar and some in the backyard.' As for those specimens on display, the situation was not much better. Because of the dilapidated condition of the building, many of the walls were supported by wooden props such that 'anyone who visits the museum ... can hardly escape the somewhat weird sensation of being in a standing forest of dead timber.' The worst feature of the building, however, was that its contents were extremely flammable – any fire would quickly reduce the wooden structure to rubble. 'Now that I have directed attention to this important matter,' Belcourt concluded, 'I trust that the present government will promptly realize its duty and ... not imitate the masterly inactivity of their predecessors in office.'[69]

By itself, Belcourt's speech placed the government in an embarrassing position. He was not the only member on the government side, however, to call for a national museum building. To dispel any suggestions that Belcourt was merely speaking in the interests of his constituency, John Charlton, the veteran MP for Norfolk North, sarcastically noted that one of the original reasons behind the removal of the Survey museum to Ottawa was the desire to locate it 'amid surroundings that were in keeping with its value.'[70] With this backbench support for the project, the ever-artful Laurier responded with great care and adroitness. He was quite willing to make available the papers that the Belcourt motion called for, but he would not commit his government to anything definite beyond advising the House that the matter was under serious consideration. 'That something should be done,' he observed, 'goes without saying.'[71]

Before the Liberals had been in office for a year, Dr Dawson had succeeded in getting the government to acknowledge that the Survey's magnificent collection needed safeguarding. In fact, within a few days of the discussion of the issue in the House, the Department of Public Works began considering suitable sites for a new museum building.[72] Despite such initial success, Dawson remained ever conscious of the need to keep after the government until the funds were approved, a site was selected, and construction was underway. Such pressure would not be easy to maintain, however, given the severe financial restrictions on the Survey. The Laurier government was determined to provide the kind of leadership that would lift the country out of its economic doldrums and set it on the road to prosperity. Perhaps more than any other cabinet minister, Clifford Sifton best personified the new administration's commitment to national development and integration. Upon assuming the Interior portfolio, he was quite anxious to reformulate the existing immigration and settlement policies to fulfil the great promise of western Canada. The Geological Survey was therefore low on Sifton's list of priorities and temporarily spared from his reforming crusade. It did nonetheless continue to suffer financially, as the government, in an attempt to reduce excess expenditure, kept the increase in the Survey budget to less than 1 per cent over the next four years.[73]

Dr Dawson was quick to realize that the Survey would be expected by the Liberals to play a fundamental role in furthering the exploitation of Canada's natural resources. He reasoned that by doing this job and doing it well, the Survey would certainly help its own case for a new building. He consequently began to emphasize the practical side of the department's field work, while at the same time trying not to lose sight of the purely scientific aspects. At the beginning of his first annual report to his new political masters, he remarked, 'as is customary, special prominence is given to facts ascertained in the course of the work which are of immediate economic importance.'[74] Dawson also tried to ensure that the Survey's field operations reflected as much as possible the government concern with developing the resources of the country. Macoun's field work over the next few years was a case in point. The Professor had planned to spend the summer of 1897 on the prairies again and had sent Spreadborough ahead to Edmonton in early April to observe the spring bird migration. Other plans for the Survey naturalist soon surfaced, however. In the spring of 1897, in order to keep the trade of the mining district in southeastern British Columbia from falling into American hands, the Laurier government negotiated an agreement with the Canadian Pacific Railway to build a line through the Crowsnest Pass from Lethbridge to Nelson. Although Macoun had explored much of this area two field seasons earlier, he was now instructed to assess the country from Fort Macleod, just west of Lethbridge, to the summit of the Crowsnest Pass, in addition to his regular natural history duties. The Professor did not mind this change in plans. He had earlier wondered whether Survey finances would allow him to take to the field at all that summer and had warned Spreadborough beforehand that he might be recalled at any time. Besides, it did not really matter where he collected as long as he was able to secure new specimens. As he optimistically told his field assistant that spring, 'We may have a big museum yet.'[75] Macoun consequently headed to Calgary in early June and spent a month working up the natural history of the nearby foothills in the company of a Dominion Lands Survey party under his son-in-law, A.O. Wheeler. He was then joined by Spreadborough, and together they clambered around the Crowsnest Pass area for the next two months. After four consecutive summers in the west, Macoun now considered his natural history survey of the region complete.

When Macoun returned to Ottawa that fall, he found the office work again in serious arrears. For the second consecutive summer his son had been dispatched to the sealing grounds, and the routine chores had perforce been neglected. The press of the Professor's various projects, however, prevented him from doing much that winter to help clear the backlog. He had spent much of his time in the foothills collecting lichens for part seven of his plant catalogue, and he was anxious to have them sent away to be examined by specialists. He also began to work seriously on an addendum to his moss catalogue – an addendum that promised to generate as much controversy as the original work, since he was still sending any new specimens to Dr

Kindberg for determination while advising Mrs Britton that he had effectively cut his ties with him.[76] Finally, Macoun copied Spreadborough's Edmonton-area field notes into his bird manuscript and began to contemplate a similar catalogue on Canadian mammals.[77]

Despite such progress in cataloguing the biological life of the Dominion, there still remained the growing problem of what to do with the specimens that Macoun and other Survey members continued to bring back from the field each year. Over the past year, in an attempt to create some additional space in the overcrowded museum display cases, Herring had remounted 346 birds on smaller wooden stands. It was a futile effort, however, for there was still no room for larger items, such as the fine seal specimens that Jim had brought back with him from the Pribilofs in 1896.[78] The storage situation was equally bad. Following Prime Minister Laurier's recognition in the House in May 1897 that the display facilities at Survey headquarters were inadequate, Dr Dawson had been given permission to rent the lower two floors of the adjoining building for storage. Yet this additional space was unsatisfactory because the building was even more liable to fire than the main building, and it was not long before an unrelenting Dawson was hammering away at the government again for more spacious, fireproof quarters for the Survey offices and museum. In his annual report for 1897, he pointed out that 'the economic and scientific value of the collections and records ... and the impossibility of replacing them if destroyed is not fully appreciated. Nor is it possible, in the present cramped quarters, to give any just exposition for the public eye, of the material wealth of Canada.'[79] Dawson also got in touch with Belcourt again, privately advising him that he should ask for a copy of the report of the Departmental Commission on Records.[80] Appointed 4 March 1897 in response to the fire in Parliament Hill's west block one month earlier, this three-man commission – composed of the deputy minister of finance, the auditor-general, and the under-secretary of state – had been formed to investigate the conditions under which records were maintained and preserved in the various government departments. When the commissioners visited the Survey on the afternoon of 11 May 1897, Dawson personally escorted them throughout the building. He revealed to them how valuable maps, plans, field books, correspondence, and other important documents, accumulated over the past half century, were either crammed into Survey offices or squirreled away in the basement or in one of the sheds in the yard. He also pointed out that the only supply of water in the building came from one small tap, while there was not even a hose or hydrant outside.

When Belcourt got his hands on a copy of the report in April 1898, it was quite evident that Dawson's efforts had not been lost on the commissioners. The report described the scandalous conditions under which the Survey's records were kept. It also drew attention to the precarious state of the museum, in particular of Macoun's herbarium, even though this aspect was not officially part of the commission's mandate. It concluded by stating that, 'the Commissioners were profoundly impressed by the urgent necessity for the adoption of prompt and effective measures to minimize the risk of fire to which the valuable contents of this department are constantly exposed, and also by the inadequacy of space and of general facilities necessary to the proper conduct of this highly important branch of the public service.'[81] Belcourt could not have asked for better ammunition in his fight for a new museum building. Rising in the House on 8 June, five days before the session ended, the Ottawa MP first read several extracts from the report and then called upon the government to take up the commissioners' recommendation and make provision for a new building in the next session's estimates.[82]

Dawson took up Belcourt's lead before Parliament resumed business that fall and, together with the new chief architect, David Ewart, called upon Sifton with sketches for a museum building in Major's Hill Park.[83] Dawson was concerned that, unless a tentative plan was adopted and estimates were drawn up, there could be no serious consideration of the project. 'I have every confidence,' he counselled Sifton several days after the meeting, 'that any action taken in the matter of affording suitable accommodation for our collection and offices will be approved by the Government and endorsed by the public.'[84] This position received considerable support from those organizations that stood to gain one way or another from a national museum building. In the *Canadian Record of Science*, the existing Survey building was described as 'nothing short of a disgrace,'[85] while a *Canadian Mining Review* editorial argued that 'the most precious collection of minerals, fossils and botanical specimens on the American continent ... are [sic] worthy of a better fate.'[86] The Royal Society of Canada took a different tack. Referring to Laurier's June 1893 pledge to make 'the city of Ottawa the centre of intellectual development of this country, and the Washington of the North,'[87] it suggested that 'the Government of the Dominion has now an admirable opportunity of taking a practical step towards giving the Washington of the North some of the aspects of the Washington of the South.'[88] No funds for a national museum, however, were contained in the new session's estimates. When Conservative Sam Hughes raised the matter in the House on 26 April 1899, Sifton simply responded, 'The matter has received a very considerable amount of attention from the Government.'[89] Something was definitely afoot though. That same day, following a tour by the minister of public works of Survey headquarters, Dr Dawson submitted a twenty-six-page memo justifying the erection of a new museum. One month later, Belcourt received assurances that funds for a museum building would be included in the supplementary estimates.[90]

While the battle for a national museum was reaching a climax, Macoun continued to perform the kind of work that, in Dawson's words, was 'of immediate value to the public from an economic standpoint.'[91] He had spent the previous summer assessing the agricultural capabilities of Cape Breton and, on the basis of the thousand plant specimens he had collected, predicted that the capabilities of the island 'had been much underrated.'[92] Now, in 1899, he was dispatched by Dawson to gather 'facts relating to the economic value or agricultural character of the soils' of New Brunswick and Sable Island 'as indicated by the plants or otherwise.'[93] Macoun found these two summers in the Maritimes, particularly his visit to Cape Breton Island with his wife and daughter Nellie, more like a holiday than work. He had quite a chuckle when a fisheries officer at Baddeck mistook him for an old tramp in search of food along the sea-shore. Life was also much happier for his son. On 14 December 1897, Dawson took up the matter of Jim's temporary status with Sifton and, on the basis of the younger Macoun's Behring fur seal work, secured his appointment to the Survey's permanent staff.[94] Six months later, not only was Jim officially named assistant naturalist, but the mandatory civil service examination was waived on Dr Dawson's recommendation that no one else in the service possessed similar professional qualifications.[95] Shortly thereafter, Jim was put in charge of the Canadian forestry display at the 1900 Paris Exhibition and took his wife overseas with him. In keeping with Dawson's desire to highlight the practical side of the Survey's endeavours in order to justify the claim for better facilities, Jim was directed to do all he could to promote Canada's forestry industry.

In the meantime, the once-bright prospect of a new Survey building had faded. When the supplementary estimates were brought before the House in July 1899, there was still no provision for the museum. It had evidently been struck out at the last moment.[96] The reason for the government's change of heart is not clear. It is apparent, nonetheless, that the discovery of gold in the Klondike had made it necessary for the Department of Public Works to spend large sums of money on public buildings, roads, trails, bridges, and telegraph lines in the Yukon.[97] Under the circumstances, the Laurier government probably decided that it would be impolitic to embark on another expensive project at this time. Dawson did not give up hope but suggested to Belcourt that 'a sum may be obtained before the end of the session if merely definitely to affirm the principle and inaugurate the work.'[98] He was to be disappointed again. When Parliament prorogued on 11 August 1899, no money for the museum had been approved. Still, there was good reason to believe that funding might be forthcoming in the next session. As B.E. Walker, president of the Canadian Institute, reminded Dawson, 'The failure of the Government to act this session is certainly very disgusting, although it is evident that they intend to do something sooner or later.'[99] This seemed to be the case, for in early October Dawson received sketches and plans for a new museum building from the Department of Public Works.[100] Already disappointed several times in the past, Dawson decided to leave nothing to chance and orchestrated the most intensive campaign to date.

That fall the government was bombarded from several fronts. In his presidential address to the Ottawa Field-Naturalists' Club, H.M. Ami, assistant curator to the Survey museum, remarked 'as a Canadian, as one who has at heart the development of our vast mineral as well as forestry and fishery resources ... our need of a National Museum ... is very deeply felt.'[101] The Royal Society of Canada made plans to send a second delegation to the prime minister in as many years. Perhaps the staunchest supporter of a national museum, however, was Walker. At the semi-centenary meeting of the Canadian Institute, he devoted his entire presidential address to the need to spend larger sums on national surveys and museums. 'We are rich enough to bear the costs with ease,' he scornfully concluded, 'but we are not intelligent enough to see our own interest in spending the money.'[102] This speech was subsequently printed as a separate pamphlet and distributed to Senate and Commons members.

Dr Dawson also tried to exert as much pressure as possible on the government. Convinced that Sifton was well aware of the problem, he set his sights on the prime minister. In May 1900, he visited Laurier twice in a two-week period, first in his capacity as Survey director and then as a member of a Royal Society delegation. His first interview was followed up by a letter sent at Laurier's request in which Dawson recounted the arguments in favour of a national museum. Describing the present Survey building as a 'mere fire trap,' he explained that although 'a handsome and therefore an expensive building would no doubt be a credit to the capital ... a serviceable building, promptly provided, would be sufficient.'[103] He also argued that the museum served as a medium for advertising the economic potential of the nation; that this potential role of the collection was not being exploited, he told Laurier, was particularly distressing. Despite such arguments, the Laurier government still did not vote any money towards a national museum building during the 1900 session. With a general election on the horizon, the Liberals were apparently concerned about how the Canadian public would respond to such a major undertaking and decided it would be politically wise to wait a few more months.

Macoun, meanwhile, provided further evidence of the kind of museum-related activity that was being carried on at the Survey with the publication of the first part of his *Catalogue of Canadian Birds*[104] in April 1900. Utilizing the American Ornithologists' Union checklist as the basis of arrangement, the professor drew freely upon all existing ornithological literature, as well as the unpublished notes of field observers, to outline the range, migration patterns, and breeding habits of the birds of Canada, Alaska, Newfoundland, Labrador, and Greenland. In a move that was probably unintentional but nonetheless telling, he also provided a list of the few Survey bird specimens that could actually be displayed in the museum. As in the case of his botanical work, the long-awaited bird catalogue was well received. J.A. Allen of the American Museum of National History, who had earlier criticized the Survey's ornithological work, described it as 'a compendium of ornithological information for the northern half of North America of great permanent interest and value.'[105] Dr Merriam, a co-founder of the AOU, was also impressed. 'The appearance of the first part of your Catalogue of Canadian Birds,' he congratulated Macoun, 'marks an epoch in the history of Canadian ornithology.'[106] Macoun's rival at the Survey, J.F. Whiteaves, felt otherwise. In an apparent fit of jealousy, Whiteaves, who normally prepared publication notices for such works, neglected to do so in the case of the new bird catalogue.[107] Such pettiness did nothing to ease the already strained relationship between the two men.

If there was anyone who deserved particular credit for the bird catalogue, it was the Professor's field assistant, William Spreadborough. Macoun had once referred to the catalogue as 'our joint work'[108] – a remark that was subsequently confirmed by the many references to Spreadborough's observations and collections. Since 1889, Spreadborough had performed most of the ornithological field work; in 1898, for example, while Macoun was botanizing on Cape Breton Island, Spreadborough went west again with another Survey party to collect birds in the Yellowhead Pass area. Those times, moreover, when he had been sent ahead in the spring, he had often used his own money until Survey funds were forthcoming. By all rights, Spreadborough was the unofficial Survey ornithologist. The quiet, unassuming backwoodsman from Bracebridge probably liked it that way. He worshipped the Macouns and willingly dropped anything to take to the field with them.

Macoun's field work in 1900 once again reflected the recent trend towards practical results. In April of that year, Ontario Premier George Ross had asked the minister of the interior whether the Survey could prepare a report on Algonquin Park. Located on the edge of the Canadian shield in the heart of Ontario, the park had been established in 1893 to protect this important watershed area, as well as to provide recreational facilities. Dr Dawson subsequently dispatched Macoun to report on the economic potential of the park. Since the Survey's knowledge of the natural history of this area was spotty, Macoun took Spreadborough along and conducted a complete survey of the western part of the park over the summer months. He subsequently warned the Premier that the park's value as a game preserve would be ruined if the destructive lumbering practices were allowed to continue.[109]

That winter, Macoun finally got ready for publication the seventh and final part of the *Catalogue of Canadian Plants*. The question of whether the magnificent Survey collection would get a new home was also resolved. On 8 February 1901, three months after the Liberals had won a convincing victory at the polls, the minister of public works sent a simple one-sentence note to his deputy: 'Veuillez mettre aux estimés supplementaires une somme de $50,000 pour la Musée Geologique.'[110] It is not clear, however, whether Dr Dawson had the satisfaction of knowing that his efforts had not been in vain. The government apparently did not bother to inform him of its decision, for on 14 February 1901, he was back at his letter writing, this time complaining to Sifton that 'the insecurity, insufficiency and hopelessness of our present building is perhaps not so fully realized by some members of the Cabinet as it might be.'[111] This pessimistic letter was followed up by another one to the minister of public works on 27 February 1901, in which Dawson reminded Tarte of 'the strong claims which I think the Geological Survey has for some consideration in the way of a better Museum and office accommodation.'[112] Three days later Dawson was dead.

The crucial first steps toward a national museum had been taken. The fact that the Survey housed a magnificent collection that not only needed safeguarding but could also be the means of promoting the resources of the country could not be denied much longer. The Laurier government had little alternative but to act; the Geological Survey's field activities, in particular its retrieval of geological, ethnological, and biological specimens, had created the need for a national museum building in Ottawa. Besides, the Liberals were enjoying the benefits of an economy that was expanding by leaps and bounds. Not only could a new museum building be afforded, but such a structure would give expression to the general spirit of optimism that accompanied this economic growth. This decision to build a new home for the Survey was a welcome relief for the natural history division. Ever since Macoun had joined the staff of the Survey, he had faced the perplexing problem of where to place his voluminous collections – a problem that grew only worse as his duties expanded. In his 1900 report, he noted that, in addition to the herbarium holdings and material already in the museum, there were more than 2,000 bird and mammal skins, 100 reptiles, and 100 freshwater fish on hand.[113] A national museum would mean that these collections would be rescued from the dark reaches of the Sussex Street building and brought before the public. New facilities would probably also result in the enlargement of the natural history staff. The Professor, his son, and William Spreadborough would no longer be responsible for performing the lion's share of the Survey's natural history work. Little did Macoun or anyone else suspect, however, that occupation of the national museum building was some ten years away.

Canada's Century

Following George Dawson's unexpected death, Clifford Sifton finally began to consider ways to make the Geological Survey of Canada a more practical institution in line with his overall development policies. The increasingly powerful minister of the interior wanted to reform the Survey in such a way that it would provide the kind of technical assistance that would facilitate expansion of the mining industry, one of the driving forces of the Canadian economy and main sources of its export income. The early 1900s were therefore reminiscent of the troubled 1880s, when the Survey was under attack for not doing enough to promote the material interests of the country. Yet, as in the case of the earlier period, except for a bitter personal controversy over James Macoun's assessment of the Peace River district, the natural history work of the Survey emerged relatively unscathed.

Much of the credit for the progress made by the natural history branch during this period must fall to Dawson's successor, Dr Robert Bell. Described by the Survey's official biographer as a 'serious humorless, duty-driven Presbyterian,'[1] Bell brought to the position an unrivalled record of service that dated back to 1865. He also less fortunately brought with him a penchant for intrigue and controversy that carried over into his five-year term at the Survey helm. What mattered to the natural history branch, however, was Dr Bell's determination to demonstrate beyond any doubt that he was the best man ever to lead the Survey. He wanted all facets of the Survey, not simply the geological side, to reach new heights of performance.[2] Professor Macoun could not have asked, then, for a better successor to

Dawson than Robert Bell. The geologist was not only an avid naturalist but, like Macoun, placed a great premium on field work and data collection. They were both all-round scientists, holdovers from the mid-nineteenth century. The pair were also strikingly similar in personality and character. They were stubborn, arrogant individuals who valued their own work above everything else and were willing to go to almost any length to see it carried out. It boggles the mind to contemplate what life at the Survey would have been like had these two men ever had a major falling out.

Dr Bell's preference for wide-ranging surveys and collection, as well as his desire to placate his political masters, meant that Professor Macoun would continue his practice of the past several years of assessing the potential of various regions on the basis of their flora – the kind of field work that had originally secured him a position on the Survey staff some twenty years earlier. Bell, in fact, took advantage of the Survey's improved financial situation to have as many parties as possible in the field each year. Macoun's son Jim consequently conducted his own separate surveys. Bell also encouraged geological parties to bring back material, as well as taking steps to ensure that the Survey would have a collection to move into the new building, the Victoria Memorial Museum.[3] The natural history branch responded to these favourable conditions by making great strides. It reached the point where the Macouns depended far less on outside assistance with their botanical work than they had in the past. The same could not be said of their faunal work, but then they were botanists first and zoologists second. Even here, though, the *Catalogue of Canadian Birds* was completed and a similar catalogue for mammals started. The Professor's enthusiasm, meanwhile, was as boundless as ever, and he looked upon each new field assignment as an opportunity to endorse the idea that the twentieth century would belong to Canada.

On succeeding Dawson as Survey director, Robert Bell finally achieved the position he believed had been unjustly denied him too many times in the past. The geologist had never had any faith in Dr Selwyn's leadership and had continually plotted to discredit him.[4] He had also been thoroughly opposed to Dawson's appointment and had subsequently complained to Sifton that he had been denied the directorship because he was a Liberal in politics and the Survey was a Conservative clique.[5] Bell's appointment as Dawson's successor in 1901, however, had nothing to do with soothing the geologist's hurt ego but was related to Sifton's proposed reorganization of the Geological Survey.

Ever since the 1884 Select Committee investigation into the Survey's activities, parliamentarians of both political stripes had regularly demanded that the government department concentrate on gathering practical scientific information that could be used by private interests. Prime Minister Macdonald apparently concurred, for when the issue arose in the House of Commons in May 1888, he stated, 'I think it is capable of improvement ... to make it more of a practical institution giving attention to such subjects as mining, instead of exclusively to the more scientific operations that have been carried on for years under Dr Selwyn.' The death of the minister responsible for the Survey, Thomas White, however, effectively stalled reform of the agency. As Macdonald confessed, 'I do not know exactly what his plans were,'[6] and it was not until 1890 that the Conservative government got around to revising the 1877 Survey Act. Even then, the modifications largely reflected Selwyn's ideas.[7] Reverting to its old name, the Geological Survey of Canada became a separate department under the supervision of the minister of the interior. The education qualifications of its technical officers were raised and a conflict-of-interest clause was included. Other than these administrative changes, the duties of the Survey remained the same, except for the added responsibility of collecting information on Canada's water resources. The new department was still to 'maintain a museum of geology and natural history,' as well as continue to 'collect, study and report on the fauna and flora of Canada.'[8]

The pressure on the government to rein in the Survey had not lessened by the time Clifford Sifton assumed the Interior portfolio; just before the Liberals came to office, Dawson had observed to a fellow Survey officer that 'the only "credit" heretofore resulting from ... northern trips has taken the form of grumbling that the government should send expeditions to such regions when districts nearer home remain to be examined.'[9] Sifton shared the concerns of many Survey critics who believed that the government agency limited its effectiveness by producing lengthy, complex reports instead of simple, concise summaries that would facilitate the rapid development of Canada's mineral wealth. The new minister, however, was initially preoccupied with the reformulation of settlement and immigration policies. He also balked at taking on Dr Dawson, who was equally concerned with upholding the Survey's reputation in the international scientific community. Survey reform was consequently not pursued with any alacrity until Dawson's death, which, although a great loss to Canadian science in general and the Survey in particular, provided Sifton with an ideal opportunity to initiate the reforms sought since the early 1880s.[10] Until the necessary legislative changes to the 1890 Survey Act were forthcoming, he needed someone to assume a caretaker role. Dr Bell, as the senior Survey officer, was the logical choice. That his selection was of a temporary nature is evidenced by the fact that he was appointed by ministerial letter and not order-in-council, and simply named acting director.[11] Perhaps E.G. Prior, Conservative MP for Victoria, best summed up Sifton's motives for appointing Bell: 'I suppose a bird in the hand is worth two in the bush.'[12]

Conscious of the Laurier government's overriding desire to exploit Canada's great resource wealth, Dr Bell sent an unprecedented number of parties to the field in 1901 to gather practical resource information.[13] Macoun was dispatched in May to see whether the fruit-growing district of southwestern Ontario could be enlarged, and he spent four months examining the plant life, particularly the trees, between Niagara Falls and Owen Sound. He later reported that 'the capacity of the whole region was only limited by the amount of intelligence brought to bear upon its natural capabilities.'[14]

This kind of positive approach to resource development had been the hallmark of Macoun's field work. Indeed, since joining the Survey, he had never drawn a distinction between pure and applied research and had always been willing to use his knowledge for utilitarian purposes. Such was the case when he appeared as an expert witness before the Senate Select Committee on the Resources of the Great Mackenzie Basin in April 1888. The brain-child of Macoun's Manitoba friend John Christian Schultz, now a Conservative senator, the committee was charged with gathering information on the exploitable resources of Canada's 'Great Reserve' north of the Saskatchewan watershed, east of the Rocky Mountains, and west of Hudson Bay. Traders, missionaries, politicians, and scientists who had either lived in or visited the region were consequently called upon to give evidence. Macoun qualified on the basis of his participation in the 1875 Selwyn expedition, but from his testimony it seemed as if he had spent the better part of his life in the region. Ready to field any question put to him, he made three separate appearances before the committee. In keeping with his recent promotion to Survey naturalist, the Professor described how he was able to determine the capabilities of a region on the basis of its flora *and* fauna. 'The character of the animals of a country,' he explained in simple terms, 'at once indicates the character of a country.'[15] He also provided a wide-ranging assessment of the economic potential of the biological life of the region, complete with species lists. The most intriguing part of his testimony, however, came during his discussion of the superior quality of the grains of the North-West when he suggested that Canada as a northern nation was destined to be a dominant society. Chairman Schultz tried to intervene at this point but to no avail. The Professor continued, 'The men on these plains will be fearless, strong and hardy ... it is a land perhaps not flowing with honey, but it is land teeming with everything that makes the heart of man glad.'[16] Despite the excesses of his testimony, Macoun had demonstrated that his knowledge of the geographical distribution of Canada's biological life had economic applications and that ongoing research along these lines, in keeping with his new position, would be beneficial to the material interests of the country. Clearly he would have little trouble satisfying Sifton's concerns about the role and purpose of the Survey.

For the first time in several years, James Macoun was also active in the field during the 1901 season. Returning from Paris in January, he was named naturalist to the Canadian International Boundary Commission and spent the summer working up the flora in the Fraser Valley around Chilliwack, British Columbia. He was assisted by William Spreadborough who, as usual, had been sent ahead to begin his faunal work early. This time, however, Spreadborough did not have to travel cross-country to assume his duties, since he had just taken up residence in nearby Victoria. Jim and Spreadborough proved an effective team in the field, amassing some 1,972 sheets of botanical specimens, 442 birds and mammals, and 300 reptiles that season.[17] In fact, while the Professor carried out a series of special assignments over the next few years, the pair performed most of the biological survey work. They also became like brothers and tended to look out for one another. As future Survey ornithologist Percy Taverner once recounted: 'At one camp in the mountains, James had gone to the post office for mail and supplies, returning through the bush late at night. Next morning William remarked to him, "You had a companion on your way back last night." James asked, "What do you mean?" "Nothing, but a cougar followed you most of the way." "How do you know?" "Oh, I was following the cougar." '[18] The bond continued to grow over the next two decades, as the two men continued to work together in the field.

Despite the fact that both Macouns spent the summer of 1901 in the field, they returned to Ottawa that fall to find the office work in reasonably good shape. With the Canadian economy going through an unparalleled period of growth, the Liberal government finally began to fund the Survey at a level more in keeping with its various responsibilities.[19] This increased appropriation resulted in, among other things, the hiring of a full-time clerical assistant, a Miss Stewart, who divided her time between the Survey library and the natural history branch. Jim greatly welcomed this additional help and relieved himself of his correspondence duties, as well as most of the routine herbarium chores, such as mounting, labelling and numbering. This assistance did not mean, however, that the winter months were any less busy. Now that father and son were conducting their own respective surveys, the amount of material that had to be processed each winter grew substantially. As the Professor finished one project, moreover, he took up others. During the winter of 1901–2, for example, Macoun no sooner completed the seventh and supposedly final part of the *Catalogue of Canadian Plants* than he began to contemplate another two parts dealing with fungi and seaweed. He also felt that the bird catalogue was far enough along that he could give more attention to the production of similar works on Canadian mammals and freshwater fish.[20] Of the two, the mammal catalogue took priority. Jim consequently wasted little time getting his Chilliwack specimens to Dr Merriam for identification. The American naturalist was glad to examine the collection, especially since it had been made along the Canadian-American border. He was somewhat perturbed nonetheless that a tentative list of the specimens was published in the *Ottawa Naturalist* before the collection had been sent to Washington for exact identification. 'This has led, as usual in such cases, to the publication of false records resulting from mis-identification,' Merriam lectured Jim, 'and in some instances ridiculous errors, such example as ... "Little *Stupid* Skunk" instead of "Little Striped Skunk," and so on.'[21]

For the 1902 field season, Jim and Spreadborough resumed their natural history work with the Boundary Commission, collecting in the region just west of the Columbia River near Trail, British Columbia. The Professor, in the meantime, was dispatched to evaluate the agricultural possibilities of the Klondike region, the farthest point north that he had ever explored in his career. The seventy-one-year-old Macoun did not leave for the Yukon until June, delaying his departure because he felt that the vegetation would not be very far advanced if he went any earlier. Yet upon his arrival he was astounded to find the local flora in bloom and producing fruit. After spending a month roaming the Dawson area, Macoun subsequently reported that the long, bright summer days and warm July temperatures combined to produce conditions suitable for the growth of vegetables and grains. 'We are quite safe in predicting,' he announced in the Survey's annual report, 'a great future for the Yukon district as a producer of everything needed to support a very large population.'[22] Macoun expounded on these findings before the Commons Committee on Agriculture and Colonization on 17 April 1903. Before doing so, however, he presented himself as the unrivalled authority on the potential of the Canadian North-West, intimating that the committee members were there to listen and learn. He informed them, 'the remarks that I am going to make to you to-day are not prophecy, they are merely deductions from actual facts, and after I am dead, and many of us are dead, my words will come truer than they are to-day.'[23] He then went on to explain how the great part of the country between Edmonton and the Klondike was destined for large-scale agricultural colonization. The Professor's enthusiasm knew no limits except the boundaries of the region itself.

Macoun's assessment of the agricultural capabilities of the Yukon district was part of Bell's larger strategy to strengthen his claim to the directorship by demonstrating that the Survey could provide practical information of immediate economic value. Such work, however, neither dissuaded Sifton from revamping the Survey nor improved Bell's standing with the minister. When the acting director raised the matter of his temporary position in the summer of 1902, Sifton advised him that he did not want his appointment permanent because he needed 'an absolutely free hand' to bring about the 'complete re-organization' of the department 'with the primary view of making the economical features of the work more prominent.'[24]

There were also problems on the museum front. On 20 May 1901, one month after the new Parliament convened, the minister of public works had finally announced the government's intention to construct a national museum building in memory of the late Queen Victoria. In tabling the chief architect's plans, J.I. Tarte had warned that the estimated half-a-million-dollar cost of the museum building could rise considerably if it were to include space for the Supreme Court, the Exchequer Court, the National Gallery, and the Fishery Exhibit.[25] The cost of the building, however, had concerned the House less than the proposed location, in Major's Hill Park, opposite the corner of Mackenzie Avenue and St Patrick Street.[26] As soon as the minister had finished his speech, the members became embroiled in a heated debate over the site before an initial appropriation of $50,000 was eventually approved. It was an ominous beginning to the project. Within a year, Tarte had been obliged to go back before the House to ask that the funds be revoted. Nothing in the interim had been done because the minister had had second thoughts about the high cost of the building. The other and by far the more important stumbling block was the question of site. 'We are not building for today,' he reflected, 'but must think of the future.'[27] Tarte's inability to come to a decision on the site was made worse by the pressure the government came under from those Ottawa citizens who lived on the east side of the Rideau Canal near the current Survey headquarters. In a firmly worded petition to the prime minister, accompanied by ten pages of signatures including that of the archbishop of Ottawa, they complained that the removal of the museum would result in a considerable loss of business and that they expected to be compensated by the locating of future government buildings in the area.[28]

The continuing delay in getting the national museum underway did not deter Macoun from pushing on with his various projects, and in April 1903 and November 1904 the *Catalogue of Canadian Birds* was completed with the publication of parts two and three. Again, the reception was favourable. The *Auk*, the official publication of the American Ornithologists' Union, referred to the work as 'the most extensive and valuable single contribution to Canadian Ornithology since the publication of the bird volume of the "Fauna boreali-Americana," seventy-five years ago.'[29] Percy Taverner, a regular supplier of notes and observations since 1899, was equally laudatory, describing it as 'a tremendous piece of work ... that reflects credit upon you and bestows a great boon upon us who are interested in geographical distribution.'[30] Taverner was 'a little chagrined,' however, to discover that some of his Muskoka sightings had been attributed to Spreadborough. Such mistakes were to be expected, though, given the sheer volume of work that the natural history branch handled, especially since there were now two parties in the field each season devoted almost exclusively to natural history. Even with part-time clerical assistance, the branch was always behind in its exchange program with other institutions – a situation that was further complicated by the fact that field collections were sometimes left unpacked for several months while the Macouns attended to other matters. Jim's 1901 and 1902 botanical specimens, for example, were left in their original field bundles until the bird catalogue was finished.[31] Taverner was actually quite lucky. Other naturalists had sent the Survey specimens only to have them misplaced somewhere amongst the material that lined the corridors of the Sussex Street headquarters.[32]

Such problems were expected to be resolved once the natural history branch moved into newer facilities where it would have more room to organize its various activities and collections properly, as well as employ additional staff. There was good reason to doubt, however, whether the promise of a new building would ever become a reality. On 23 March 1903, James Sutherland, the new minister of public works, asked the House to vote the initial appropriation for the Victoria Memorial Museum yet again. Like his predecessor, Sutherland could not decide upon a suitable site for the building. Nor did he give the House any indication of when a decision might be forthcoming, although he did hint that the matter was under serious consideration. Similar responses were forthcoming in July and September. Finally on 17 October, seven days before the session prorogued, Sutherland asked that $100,000 be approved for the purchase of land for the museum. That the minister's motion was approved with relatively little debate probably reflected Parliament's sense of relief that the site question had at last been resolved.[33]

In order to secure the land at a reasonable price, Sutherland refused to divulge the proposed site for the museum until after the sale had been finalized. Thus it was not until 24 March 1904, three years after the decision to erect a new building had been made, that the government announced its purchase for $73,500 of the Stewart property or 'Appin Place,' a nine-acre block of undivided land one mile south of the Parliament buildings, at the foot of Metcalfe Street. This site was evidently selected by Public Works over other possible locations in the interests of creating a grand promenade along Metcalfe Street between Parliament Hill and the new museum, similar to that which existed in Washington along Pennsylvania Avenue between the White House and Capitol Hill. In fact, chief architect David Ewart's original plans for the museum included features that complemented those of the centre block of the Parliament buildings.[34] Confident that the government had fulfilled Laurier's 'Washington of the North' promise, a triumphant Sutherland proclaimed to the House, 'Probably not another lot of land could be secured anywhere in the city so well suited for this national building.'[35]

The thought of a new building undoubtedly pleased Dr Bell and his staff. Like Selwyn and Dawson, the acting director had faced severe problems of overcrowding and had been forced to transfer several offices to the adjoining Baskerville annex.[36] Unlike his two predecessors, however, Bell now had the satisfaction of knowing that the Survey's days of operation out of the decrepit old building on Sussex Street would soon be over. In the meantime he took steps to ensure that there would be a collection to move into the new building and, within a week of Sutherland's announcement, asked Sifton for permission to post three firemen around the clock at Survey headquarters.[37]

Dr Bell did not fare as well in his own battle to have his appointment made permanent. When the temporary nature of his position was raised in Supply on 17 July 1903, Sifton explained to the House that any change in Bell's status would not be considered until after he had introduced long overdue legislative changes to the 1890 Survey Act.[38] A more scathing assessment of the situation was given to Montreal MP Robert Bickerdike six months later. Sifton lamented: 'the mind is scarcely capable of comprehending the amount of scheming, mining and countermining that goes on within the walls of the Geological Survey. In fact it is so bad that it is almost impossible for me to pay attention to the representations which are received and it is most difficult to get anything like a fair estimate made of the work done by any particular member of the Survey ... It is my intention if I am spared to keep charge of the Department for any length of time to make a radical reorganization in which the economic value of the Survey will be more apparent than it is now.'[39] This low regard for the Survey under Bell was to worsen thanks to James Macoun's 1903 assessment of the Peace River country.

The 1903 field season represented a change of pace for both Macouns. Although he had lived in Ottawa for more than twenty years, the Professor had never conducted an extensive natural history survey of the region. Any examination was largely restricted to his outings with the Ottawa Field- Naturalists' Club. He consequently decided to spend the summer at home, making day trips to local points of interest. Jim also performed work of a different nature when Dr Bell orally instructed him to assess the agricultural prospects of the Upper Peace River district on the basis of its vegetation. This was not the first time the region had been examined. Beginning with the senior Macoun's survey in 1872, the Peace River's great potential had been regularly proclaimed by a succession of government scientists – including two former Survey directors, Selwyn and Dawson. Jim, on the other hand, was an accomplished botanist, not an agricultural specialist. Although he was quite familiar with his father's practice of judging the potential of a region from an examination of its flora, Jim had performed such work only once in his Survey career; in 1888, in response to the Senate Select Committee hearings on the resources of the Great Mackenzie Basin he had accompanied the Thomas Fawcett expedition to the Churchill and Athabasca Rivers. When it was subsequently learned that he did not share his father's enthusiasm for the northland, he had apparently been muzzled by the Macdonald administration at the request of Senator John Schultz and Manitoba Premier Thomas Greenway, and a formal report of his findings never appeared.[40] In fact, it is quite possible that a request by Macoun in November 1888 that his son be appointed to the Survey permanent list as assistant naturalist was turned down because of Schultz's opposition.[41] Bell was now asking Jim to carry out a similarly sensitive assignment. The recently launched Grand Trunk Pacific Railway intended to build across the northern prairie to Edmonton and then up through the

Peace River district. A negative report on the country would not be welcome, particularly since the new transcontinental line had the backing of the Laurier government and had been partly justified on the basis of the potential agricultural worth of the Peace River valley.

From Jim's point of view, the Peace River assignment offered him a chance to emerge from his father's shadow and demonstrate his capabilities. The assistant naturalist consequently tackled the assignment with a great sense of purpose. Starting from Lesser Slave Lake on 2 June 1903, he travelled by foot to Peace River Landing and spent the next two months on the north side of the river in the present-day Fairview region. He then crossed back over the Peace at Dunvegan and examined the Spirit River, Pouce Coupé, Grande Prairie, and Smoky River regions. Despite the cold, wet weather that summer, Jim made every effort to see every acre of cultivated land and to interview as many residents as possible. He also made a careful study of the vegetation and the soil content and depth. He even recorded the minimum nightly temperature at every campsite. The result of this three-month survey was a wealth of data about the region. In keeping with Dr Bell's desire to publicize the practical nature of the Survey's work in order to strengthen his claim to the directorship, Jim's report was published in late March 1904 as a forty-eight-page pamphlet.

Given the poor growing conditions in the Peace River country during the summer of 1903 and the method of assessing the land, it was not surprising that Jim formed a rather bleak mental picture of the region's capabilities. These findings, however, represented only one season's experience and were an insufficient basis upon which to judge the region. One of the peculiar features of the western interior is the varying climatic conditions from year to year and from place to place. In the long run, it took several decades of farming experience to bring most of the Peace River country under cultivation.[42] Jim, nonetheless, was expected to pronounce on the region's capabilities when he returned to Survey headquarters – after all, his father had been doing this kind of work for decades. He also believed, on the basis of his field work, that the land had always been referred to in exaggerated terms and that it was his duty to provide a more realistic picture. This theme dominated the report's opening remarks, where Jim observed that 'nearly all the reports on the climate of the Peace River country and the fertility of the soil have been based on observations made in the valley' and 'to attribute to the whole country the climate of the valley creates a false impression.' He then pointed out how these false impressions about the country have 'already brought not a little hardship and suffering upon those who have settled in less favourable localities.' The main body of the pamphlet expanded upon these findings, describing in detail how the shallow soil, high elevation, and frequency of frosts made for questionable wheat-growing conditions, even in the re-nowned Grande Prairie and Spirit River districts. The most damaging remarks were saved for the conclusion: 'While the country that has been described should, in the opinion of the writer, not be settled by either the rancher or the grower of wheat until there is more satisfactory evidence that it is suited for either of these pursuits, it may be safely prophesied that after the railways have been built there will be only a small part of it that will not afford homes for hardy northern people who never having had much will be satisfied with very little. It is emphatically a poor man's country ...'[43]

Reaction to the pamphlet was swift. On 2 April 1904, Clifford Sifton asked Dr Bell for a copy of the instructions that he had given to the younger Macoun, while at the same time advising him that in future all Survey parties would require ministerial approval before being sent to the field.[44] Three days later, a sheepish Bell informed Sifton that his instructions had only been given orally. He did try to mollify Sifton, however, by advising him that Jim did not think his report would deter settlement of the North-West and that sending immigrants into the region under false impressions would do more harm.[45] Sifton, for his part, did not regard the matter as being at all serious. He confessed to a Manitoba friend, 'When it is sifted down it does not amount to much more than the statement that there are summer frosts. Summer frosts as we know have been the bugbear in almost every part of Canada.'[46] What he considered extremely imprudent, however, was the way and the form in which the report appeared: 'though possibly accurate in details [it] conveys an entirely erroneous general impression and can be made use of to the detriment of Canada.'[47] He was afraid that in the wrong hands it could be used to criticize national development policies. Sifton therefore instructed Dr Bell: 'I would be glad if you would take steps to see that in cases where the suitability of large tracts of country is in question the preparation of reports upon them should be the subject of consultation with me before anything else is done.'[48] Bell concurred, promising the minister that 'I shall take great care that nothing of the kind occurs again.'[49]

On 8 April 1904, the same day on which Sifton warned Bell about the potential damage the pamphlet could cause to western immigration and railway construction, the Commons Committee on Agriculture and Colonization began consideration of the matter. This was not the first time the committee had examined the work of a government scientist. For several decades now, Professor Macoun, among others, had made regular appearances before the committee whenever his expertise was sought on a particular area or issue. Nor was it the first time a Survey officer was the centre of a controversy. In *Manitoba and the North-West Frauds*, an 1883 pamphlet dismissed as the ramblings of a senile old man, former explorer Henry Youle Hind had charged that Macoun's pronouncements on the settlement potential of western Canada were nothing short of criminal – 'the concoctions of a scientific rogue'[50] – and that the government, in accepting and propagating his statements, was an accomplice. What was unusual about the committee proceedings, however, was that rather than calling upon the report's author to give evidence, the committee asked his father to appear before it. Undoubtedly flattered, Professor Macoun gave one of his better performances. Although he had not been in the Peace River country since 1875, he spoke in glowing terms of the region's agricultural potential: 'It is the most beautiful country you ever looked on.'[51] When questioned about the obvious discrepancies between his remarks and his son's report, the older Macoun stated that he did not think there was any conflict in their respective positions. 'We are both honest men,' he said, 'and both talk of what we know.'[52] Such remarks tended only to confuse the committee members, and it was eventually decided that the assistant naturalist should be the one in the witness chair. The Professor's testimony was consequently cut short, but not before many doubts about the reliability of Jim's report had been raised in the minds of the committee. In fact, Macoun's evidence probably made his son's subsequent reception all the more heated. As he told the committee at one point, '[M]y son is of age, and I am not answerable for him, and if you do not agree with him, heckle him all you can.'[53]

Jim made the first of nine appearances before the committee on the morning of 14 April 1904. Some kind of fracas was anticipated for, three days earlier in the House, a Liberal member of the committee, Saskatchewan MP Thomas Davis, had demanded to know what department Jim worked for, his date of appointment, his duties, and his salary.[54] Frank Oliver, the Independent Liberal MP for Alberta and editor of the *Edmonton Daily Bulletin*, was even more upset. A passionate spokesman for the western farming population and the interests of Edmonton, Oliver had always been something of a thorn in the government's side, forever criticizing the Department of the Interior, in particular Sifton's immigration policies. Yet because of Oliver's strength in his riding, Sifton had little choice but to support him, albeit reluctantly.[55] During the 1900 election, he called upon his brother to try to restrain Oliver: 'Ask him to be sensible for the next three months even if he has to be otherwise afterwards.'[56]

The publication of the Peace River report was something of a red flag to Oliver. In late March 1904, he had written to Sifton about possible improvements to the existing trail between Edmonton and Peace River. He justified the expenditure on the grounds that 'the sooner people can get into the country the better it will be for the Railway when built, and the greater justification for the policy of building it.'[57] What Oliver forgot to add but probably had in mind is that an improved trail would speed up construction of the rail line and open the potentially lucrative trade of the region to Edmonton sooner. With the release of Jim's pamphlet, however, not only had the Peace River country been slighted, but the wisdom of sending the Grand Trunk Pacific Railway through the region, ironically one of the few Liberal western policies that Oliver endorsed,[58] was thrown into doubt. The report also appeared to lend credibility to the Alberta MP's position on the Geological Survey. Unlike other Survey detractors who believed that the government department should simply focus on practical problems, Oliver strongly doubted whether Survey officers could handle work of this nature.[59] The Peace River pamphlet was a case in point.

The feisty Oliver consequently decided to use the committee hearings to prove that James Macoun was 'dead wrong.'[60] In the meantime, he launched a concerted campaign in the *Bulletin* to undermine the assistant naturalist's credibility. 'And after Canadians have been satisfied for one hundred years ... that the Peace River was one of the country's most valuable assets,' a 6 April editorial sarcastically noted, 'comes forward a young government employee, who on the strength of a visit of a few months, and with all the authority which official standing can give behind him, proclaims loudly and widely that it is all a mistake.' It then touched upon the glaring discrepancies in the assessments of father and son – 'the elder could see over the fence, while the younger could see only the fence' – before citing the detrimental effects that Jim's 'absolutely erroneous' conclusions might have on the Grand Trunk Pacific project. The editorial closed with a call for Sifton to dispense with Jim's services: 'The idea that a man of only theoretical knowledge and ideas should be allowed to condemn ... the Peace River country ... is not to be submitted to. Mr. Macoun is not to be condemned because he is intentionally dishonest, although he may be, but simply because he does not know, and the country cannot afford to pay men to talk, when they do not know.'[61]

Two days later, the *Bulletin* carried an interview with Albert Tait, a former Hudson's Bay Company Peace River employee who had conferred with Jim on his way through Edmonton. According to Tait, the younger Macoun 'was already possessed of an unfavourable impression of the Peace River district' – an impression that was enforced by the unseasonal weather that summer. Besides, what else could be expected from 'an amateur traveller,' 'a homesick kid,' 'a young man unfamiliar with western conditions.'[62] J.K. Cornwall of Lesser Slave Lake attributed a more calculating purpose to the report, suggesting on 12 April that Jim 'wanted to get some notoriety from giving "a black eye" to a country which was now occupying public attention.'[63]

The Conservative Opposition was also quite interested in the Peace River pamphlet but for entirely different reasons. At the time the report appeared, Parliament had just begun to debate a number of government-introduced amendments to the original 1903 Grand Trunk Pacific agreement. In responding to Laurier's proposed changes on 5 April 1904, Conservative leader Robert Borden objected to the bill on several grounds, including the viability of routing the transcontinental line through northern Quebec and Ontario.[64] Jim's pamphlet, however, now threw into question the Peace River section of the railway. The Conservative committee members consequently flocked to the hearings in unprecedented numbers and attempted to have Jim enlarge upon his remarks in his report. When it became evident that Oliver was equally determined to discredit him, they went to the extreme length of trying to pass a motion to the effect that the committee hearings go unrecorded and that Jim's report be allowed to stand on its own.[65] Such support did not help Jim's situation, but led to the conclusion, as Oliver's *Bulletin* repeatedly charged, that the assistant naturalist was in the pocket of the Conservatives.[66]

Caught in the middle, James Macoun was equally determined to hold his ground. Having already resigned himself to the fact that he would probably lose his job over the affair, he did not care what he said or to whom.[67] The chemistry was therefore right for one of the most explosive committee hearings in parliamentary history. Oliver set out to show that the assistant naturalist had but a limited practical knowledge of the region and that existing agricultural activity contradicted his conclusions. Jim, on the other hand, boldly claimed that he had tried to include in his report everything favourable about the country that he could find in previous assessments and defied the committee members to give any reputable authority that would disprove his contentions. It was not long before the two men locked horns:

Q. [Oliver] ... I believe that he has deliberately and purposefully caused inferences to be drawn which are absolutely and utterly misleading and injurious to the last degree to the best interests of this country?

A. You will allow me to take a few minutes. I do not know this gentleman's name –

Q. My name is Oliver. I represent Alberta.

A. I am glad to know you, I have heard of you before. In all of what Mr. Oliver has said he is talking to the gallery, as you all know.

Q. I ask the chairman if that is a proper thing for an officer of the government to say?

A. I do not care from whom he gets his opinion. I have worked for twenty-three years for the Canadian government.

[Mr. Davis] Too long. (Cries of 'No,' 'Shame,' 'Contemptible.')[68]

The session adjourned shortly thereafter, but not until fellow committee member Sam Hughes made the two men shake hands.

After such a raucous beginning, the committee did not resume its examination of Jim until eight days later, and then it held four meetings in the span of a week to make up for lost time. Any hope that Oliver would have cooled down in the interim, however, was quickly dashed. Still smarting from the first day of hearings, Oliver picked up where he had left off and went after the witness with a vengeance. In an effort to demonstrate that Jim simply did not know what he was talking about, he insisted that he provide hard facts not opinion and then refused to accept his answers. As a consequence, the pair almost came to blows again on April 29.

A. [Macoun] ... He is trying to make it appear that they contradict me. We will fight it out and we will fight it in the newspapers if necessary. The people will know there is no difference of opinion ...

Q. [Oliver] You are better in the newspapers than in the committee.

A. I will take my chances in both places.

Q. You are a long-range fighter?

A. I am close to the Committee now.[69]

The threat of a nasty public battle over the report was too much for Sifton, and on that same day he instructed Bell to discontinue the pamphlet's circulation. The minister also talked privately with Jim and told him what he thought of the pamphlet and that it was not to be distributed.[70]

Oliver, in the meantime, not only continued to badger the witness but became quite flippant. This exchange on May 3 was typical:

Q. [Oliver] ... we cannot accept your statement on this point as being absolute anymore than any other?

A. In what way was I mistaken?

Q. Oh, never never mind ...

A. But I would like –

Q. Oh, take my word for it –[71]

Such tactics did not ingratiate Oliver with most of the other committee members, Liberal or Conservative. They were embarrassed by his incessant contradiction of the witness and argued that Jim should be allowed to give his evidence freely and without interruption. They also resented Oliver's monopolizing of the hearings. Yet whenever they voiced these concerns or came to the assistant naturalist's defence,

Oliver turned on them, complaining to the committee chairman that he was being persecuted. By 6 May, the eighth session, the hearings had become, in the words of one committee member, 'a bear garden,'[72] with Oliver bickering with his fellow examiners almost as much as with the witness. Four days later, several of the committee members took it upon themselves to end the circus and referred the matter to a sub-committee for consideration. Jim, for his part, was forbidden to discuss his report further on pain of dismissal.[73]

The controversy was far from over. The Conservative Opposition, thoroughly disgusted with Oliver's behaviour on the committee, decided to force him to fulfil his pledge during the hearings to repeat his charges on the floor of the Commons.[74] When the vote on the salaries of the Geological Survey came before the House on 20 July 1904, Conservative MP Andrew Ingram asked why James Macoun had not been dismissed in light of the fact that a member of the government side had accused him of incompetence. Torn between supporting the administration's railway plans and defending a respected government scientist, Sifton responded: 'All that I have to say is this, that the conclusion I arrived at with regard to Mr. Macoun was that while perfectly competent, so far as scientific attainments are concerned, he drew sweeping conclusions from rather insufficient data.'[75] The Opposition was not satisfied. Thomas Sproule, the veteran parliamentarian for Grey East, confessed that he was willing to vote the assistant naturalist's salary but asked Sifton to make it clear whether Jim was guilty of deliberately making a false report. 'We do not wish to embarrass the government,' he said slyly, 'but we want to know what they are going to do about the matter in view of what occurred in the committee of this House.'[76] At this point, the prime minister stepped in and dismissed the affair as an error in judgment: 'I do not believe that he made an incorrect report with intention. I think perhaps he acted with some levity, was not careful enough, but even if he acted without sufficient accuracy, and acted too hastily in placing before the Canadian public a report in contradiction to all the accepted doctrines, even believing all that, I would not dispense with Mr. Macoun's services. I think he is a good man.'[77]

Frank Oliver had been strangely silent during the entire debate in the House about Jim's report. It is possible that he had been told by Sifton or Laurier to keep quiet on the issue. The government could not keep the vitriolic backbencher silenced for long, though, particularly since the Opposition was gunning for him. The showdown came the next day during a lengthy late-night debate on immigration literature when a Conservative member of the committee derided Oliver's behaviour during the hearings, especially his threat to horsewhip the witness. Oliver was immediately on his feet to deny the charge, declaring, 'I said that James Macoun deserved horse-whipping, and I say so still. But I did not say that either I or anybody else was going to horse-whip him.' Someone on the other side then cracked, 'You knew better than to try it.'[78] This remark caused Oliver to repeat his charges against Jim's report, but before he got much further the deputy speaker ruled that any discussion of the committee proceedings was out of order until it had released its report.

The final word on the Peace River pamphlet was reserved for the Commons Committee on Agriculture and Colonization, which issued its report on 3 August 1904. The committee pointed out that Jim was virtually alone in his stand on the agricultural capabilities of the region; even his father disagreed with him. It also stated that the printing of the report in pamphlet form was regrettable and that it should be suppressed until the country had been re-examined. What the committee was most critical of, however, were the sweeping conclusions the assistant naturalist had made on what it regarded as insufficient data: 'Mr. Macoun spent less than three months in the Peace River country, ... and it is unreasonable to suppose that any man could, within that time, acquire sufficient knowledge to enable him to make the report and give the evidence which Mr. Macoun did.'[79] Ironically, Professor Macoun had been guilty of making such sweeping judgments on the basis of one season's work for the past thirty years. Just a year earlier he had acclaimed the agricultural potential of the Yukon before the same committee, and no one had questioned how he could make such statements after only one month in the field. The contradictory experience of the two Macouns before the same committee vividly demonstrated that science was to be the servant of national development – the key that would unlock the resource wealth of Canada. Positive assessments of the country's potential were preferred, no matter how limited the data nor how bold the generalizations. Negative reports, despite being comparable to other assessments in terms of method and observation period, were frowned upon.

Curiously, when the new provinces of Saskatchewan and Alberta were created the following year, the northern boundaries were set at the sixtieth parallel of latitude, several degrees above the Peace River district, in the belief that this northern district was suitable for agricultural settlement.[80]

The fallout from the controversy was relatively restricted. In addition to receiving rough treatment at the hands of Oliver and having his reliability questioned in the Commons, Jim was prevented from resuming his work along the international boundary; Dr Bell did not want him to wander too far from Ottawa that summer. It was a small rebuke, especially since the assistant naturalist could have lost his position. Frank Oliver was largely responsible for this outcome. His disgraceful conduct during the affair had caused many to sympathize with Jim. Sam Hughes, for example, although disagreeing with the report, told the House, 'I think an officer who has the energy, pluck and manhood that Professor [sic] Macoun had should be continued in the department and that he should receive a good increase in salary.'[81]

William Spreadborough was also affected to a limited extent. It had been informally arranged during the winter of 1903–4 for the field assistant to accompany a Dominion lands survey expedition to the south and west coast of Hudson Bay. Spreadborough, however, had taken part in the controversial 1903 survey of the Upper Peace River country, and it was not until the end of May 1904 that he was finally endorsed as the expedition's naturalist. In outlining Spreadborough's duties for the season, Jim remarked, 'You have no idea how much trouble there was in getting you to work at all this spring.'[82]

The only other immediate victim of Jim's Peace River report was the Survey's acting director. During the 20 July Commons debate on the question of the assistant naturalist's competency, Sifton made it clear that Dr Bell had erred in printing Jim's report in pamphlet form – the very same criticism that the Commons Committee on Agriculture and Colonization subsequently levelled. The minister was also miffed by Bell's failure to issue his instructions in writing.[83] As a result, the chances that Bell's position would be made permanent were virtually nil as of July 1904. Sifton made his feelings quite plain the day before when he was being peppered in the House with questions about Bell's situation. 'If I remain in office and endeavour to put the Geological

Survey on a much more useful basis,' he pledged, 'I do not think that he is a good man for the position and would not recommend him.'[84]

While the controversy swirled around his son's report, the Professor had left for Banff National Park following his own appearance before the committee. Macoun had already explored much of this region in 1891 and 1897, but with the enlargement of the park's boundaries the government had decided that the collection of the local museum should be revised and updated. Macoun's son-in-law A.O. Wheeler, moreover, would be surveying in the area that summer, and it made good sense to have the naturalist work out of the same camp. Macoun consequently spent the summer collecting in the vicinity of the Kicking Horse Pass, paying particular attention to the mammal life of the region for the catalogue he was planning on the subject. Unlike his wide-ranging general surveys of the past, the Professor's investigations had now become quite narrowly focused. 'I seem to have lost interest in anything beyond what I am at present interested in,' he confided to an old Maritime colleague, 'and hence leave matters of general interest to younger men.'[85] This did not mean, however, that he was slowing down. After spending the summer clambering about in the Rockies, he was back in Ottawa for only a few months before he slipped south to consult the collections in Washington. It was a highly successful visit. Botanist Edward Greene of the United States National Museum reported back to Jim: 'All are surprised, as well as pleased at your father's vigour and enthusiasm and at the want of any signs of age about him ... one learns so much, and so many interesting things from him.'[86]

With the Professor caught up in his special projects, the general botanical work of the natural history branch became more than ever the exclusive domain of his son. Here significant progress was being made as well. In the Survey summary report for 1904, Jim noted that the former practice of sending their field collections to botanical specialists had been discontinued. His growing expertise, combined with the botanical literature at his disposal, now enabled him to work up the bulk of their collections. Any outside assistance was sought only to determine doubtful specimens or describe new species. 'There is only one cook as far as Canadian plants are concerned,' he would later defend his work, 'and that cook is myself with my father's assistance.'[87] This coming of age actually meant more work for Jim. Not only was he handling an increasing amount of the material that would normally have been sent away to specialists but, in order to keep up with the constant changes in the discipline, he had to re-examine and rename the herbarium holdings as soon as new monographs and revisions were published. The number of duplicates that the natural history branch was able to send out each year was thus limited. Other herbaria, however, were quite willing to wait, for a parcel of Survey duplicates invariably contained many rare and interesting forms.[88]

The 1905 field season marked a return to normal following the troubles of the year before. Jim Macoun and William Spreadborough resumed their natural history studies along the international boundary in British Columbia, collecting first in the little known Osoyoos Lake region before moving westward to the Skagit River. The senior Macoun, on the other hand, jumped from project to project. He carried out a study of the climatic conditions of the St Lawrence Valley below Quebec City and then gathered more than seven hundred species of fungi in the Ottawa region. When the weather eventually forced him to move indoors, he began work on his catalogue on the mammals of Canada. Much of the data for this book had been amassed by Spreadborough over the past few field seasons. In the spring of 1904, for example, while the matter of his trip to Hudson Bay was still in limbo, the field assistant had been busy collecting small mammals in southwestern British Columbia near Fernie and Elko.

Unlike his son, Macoun continued to rely on Dr Merriam and his assistants at the bureau of biological survey in Washington to examine and identify his mammal specimens. Given his own interest in faunal distribution, Merriam was pleased to learn that Macoun had commenced work on a catalogue of Canadian mammals and attended to the steady flow of parcels and boxes from Ottawa as quickly as possible. He soon became irritated, however, with the third-rate quality of the specimens – many skins were still not accompanied by skulls, and specimen labels often lacked essential field data – and he complained to Macoun in a sharply worded note that 'your collector might take a little more pains in labelling the specimens and also in making up the skins ... It costs no more and takes but a trifle more time to make first class skins.'[89] In response, Macoun defended the branch's work by explaining that the Survey's natural history duties were largely upheld by three men –he, his son, and Spreadborough –and that their total field expenses for 1905 were $769. 'Keeping this in mind,' the Professor queried, 'I would like to ask you what more could you expect of us than what you find.'[90] Merriam was impressed. 'We were all very much entertained by your graphic account of the meager facilities, and still more meager appropriation available for your work,' he acknowledged. 'It is hard for us to understand how you accomplish so much.'[91]

By the spring of 1906 it appeared that the natural history branch would not be operating from inadequate facilities much longer. Following its announcement in March 1904 that land for the new museum building had been purchased, the Laurier government called for tenders and subsequently awarded the $950,000 contract late in the year to Ottawa builder George Goodwin. Under the terms of the agreement, he was to build, by 1 December 1907, a four-storey stone building, complete with wings at each end and a fifty-eight-foot central tower; the outside walls were to be constructed of the same Nepean stone that had been used for the Parliament buildings. 'We have allowed room for storing everything that we possess at the present time,' C.S. Hyman, the new minister of public works, told the House during a briefing session in February 1905, 'and room for practically as much more.'[92] So much display space was contemplated that the building was to house the Department of Fisheries Museum and the National Gallery as well. While construction got underway, Dr Bell took steps to improve the current museum holdings and spent $6,589.10 on specimens in 1905, as compared with $62.50 the year before.[93] He also got in touch with the British Museum in February 1905 for help in drafting a set of administrative guidelines for the new museum. The elusive national museum building was to become a reality after all.

The issue of Bell's temporary status was also finally resolved at this time. At first, it appeared as though the acting director might get his wish, for on 28 February 1905, Sifton resigned his portfolio because of differences with Laurier over the education clauses of the autonomy bills creating the new provinces of Alberta and Saskatchewan. The Opposition, in the meantime, seemed to be on Bell's side. On 12 July 1905, Conservative leader Robert Borden asked the government why Bell's position was still temporary after five years; if he was unsatisfactory, he should have been replaced long ago.[94] Following Borden's lead, the acting director decided to put his case directly before Laurier ten days later: 'Sir, the torture I am suffering owing to this long continued suspense is awful. It continues to become more and more aggravated the longer it is drawn out. Therefore, I implore you and the rest of the Cabinet to take a few minutes and decide this long delayed matter.'[95] Laurier, however, wanted nothing to do with the issue and simply informed Bell, 'The matter is exclusively a departmental one.'[96] The decision on Bell's fate consequently fell to Sifton's successor, Frank Oliver.

If the chances that the acting director's position would be made permanent under Sifton were remote, they were non-existent under Oliver. Whereas the former minister could point to Bell's continual feuding with other Survey officers, his replacement had a grudge to settle with the Survey – namely the Peace River pamphlet. Oliver's dispute was really with James Macoun, but the new minister could not dismiss him, particularly after the prime minister had spoken in his favour in the Commons. Besides, although the assistant naturalist had prepared the report, it was Dr Bell who had been ultimately responsible for deciding what became of it. The need to decide what should be done about the Survey directorship therefore presented Oliver with the ideal opportunity to exact some measure of revenge for the pessimistic assessment of the Peace River country. Indeed, it is quite possible that the new minister of the interior had Jim's Peace River pamphlet in mind when he advised Laurier of his decision on 13 March 1906: 'I regret to say that while I have every confidence in Dr. Bell as a scientist of the highest class, ... my judgement is that he does not possess that administrative ability which is necessary in order that good work may be secured from the work of subordinates.'[97] Laurier, despite his earlier reluctance to get involved, broke the news to Bell in his office nine days later. This meeting was followed up by a letter in which Oliver advised Bell that the new director, Albert Peter Low, would take over at the end of the month. The acting director no longer had to worry about 'fighting off the hungry wolves who are jumping for the prize.'[98]

Dr Bell's removal from the Survey helm must have been something of a blow to the natural history branch. Even though life at the Survey under the geologist had been stormy, he was one of the best friends the Macouns could have had. His genuine interest in natural history, coupled with his desire to see the Survey excel in all areas of its mandate, meant that the branch not only survived but made great strides at a time when there was growing pressure for the kind of practical information that would facilitate the development of Canada's mining resources. In the field, while Jim headed his own collecting party, the Professor performed a series of special assignments that reflected the Laurier government's unbridled optimism in the future of the country. Back in the office, the *Catalogue of Canadian Birds* was completed, and a similar publication on mammals was underway. What would follow was a different matter altogether. Although the new museum building under construction would resolve many of the natural history branch's problems, there was a new minister in charge of the Survey. And if Oliver could not get rid of James Macoun, he could at least make life miserable for him and the branch. Operating on the assumption that 'men of the highest scientific attainments have least practical ability,'[99] he was determined to reform the Survey much along the lines that Sifton had had in mind. The future seemed uncertain at best.

The Natural History Division

The potentially difficult situation faced by the natural history branch at the time of Frank Oliver's assumption of the Interior portfolio was nothing new. The seventy-five-year-old John Macoun had known similar challenges since 1882 and had managed by one means or another to see that his natural history work did not suffer. The trouble that was anticipated with a reformed Survey under Oliver, however, never did materialize. Although the Geological Survey ceased to function as a separate government department and became a branch of the new Department of Mines in 1907, not only were its traditional, more scientific functions confirmed, but also it was no longer the responsibility of the minister of the interior. These legislative changes were followed by the replacement of Low as Survey director by Reginald Walter Brock, a veteran geologist who wanted to end the wide-ranging, all-inclusive field activity of the past in favour of specialized surveys by more highly qualified personnel. His goal was nothing less than to transform the Survey into a modern, twentieth-century institution on the leading edge of scientific investigation. He was also determined to ensure that the Survey retained control of the new building, the Victoria Memorial Museum, and he actively encouraged all areas of Survey inquiry, including natural history, anthropology, and archaeology.[1]

Far from being threatened, then, John and James Macoun had never known such favourable conditions under which to pursue their natural history work. The natural history branch was in no position, however, to become part of the leading scientific force that Brock envisaged. As the Victoria Memorial Museum neared completion, it was found that the natural history collections, apart from the holdings of the Dominion herbarium, were far from complete; moreover, the zoological specimens on hand were poorly preserved and documented. It was also apparent that the Macouns – both father and son – could not do the kind of detailed, specialized study into the life history and physiology of organisms that was underway at North American universities. Their continued preoccupation with general collecting, observation, and description was out of step with the new science of biology and its emphasis on function as well as form. 'The time had come to exploit the naturalists' wealth of description,' as one expert on the history of biology has observed, 'not perpetually to expand it.'[2] Thus, although the natural history branch experienced staff additions and was reorganized to form the new natural history division, its activities required significant updating –not a particularly easy task with Macoun still deeply committed to the nineteenth- century tradition of natural history. For the time being, the Survey was initially preoccupied with trying to make its natural history collections as representative as possible, and with sorting and evaluating the specimens that had been gathered by Macoun, his son, and other officers over the past quarter century.

When Albert Peter Low took up the reins of the Survey directorship in April 1906, it was not clear whether the needs and interests of the natural history branch would continue to receive the kind of support that had been forthcoming during Dr Bell's tenure. On the one hand, Low was quite familiar with the work of the two Macouns. He and the Professor had been appointed the same year and had spent part of their first field season together in the Shickshock Mountains of the Gaspé Peninsula. Three years later, Jim had acted as Low's assistant during his 1885 survey of the little-known Lake Mistassini region of central Quebec. The new director had also been willing to bring natural history items back from the field, particularly during his famous 1903–4 expedition to the eastern Arctic aboard the *Neptune*. At the same time, Low had been selected by the Laurier government to transform the Survey into an agency more in keeping with the needs of the time, and he had been in office only a few days before he was asked to help draft a new Survey bill.[3] Low himself was under no illusion about the reasons behind his appointment. In thanking Oliver for the honour, he wrote: 'allow me to express my desire and hope that I may fill the office ... for the benefit of the mining interests of the Dominion.'[4]

What would be the eventual outcome of this revision of the 1890 Survey Act would not be known until the legislation was tabled in the House. In the meantime, it was business as usual for the natural history branch. For the 1906 field season, William Spreadborough returned to the Chilliwack Valley of southern British Columbia and resumed his collection of birds and small mammals from the year before. Jim did not join him, as he had in past years, but remained in Ottawa for the summer to attend to routine office work. Whether the assistant naturalist was being deliberately kept from the field by Oliver because of his Peace River report is debatable. There was certainly a considerable amount of work to keep him at Survey headquarters. In addition to his usual herbarium chores, there was the preparation of the catalogue of Canadian mammals, as well as a revised edition of the bird catalogue. There were also close to a thousand pieces of correspondence annually that needed answering.[5] It therefore made good sense to have someone working in Ottawa year-round. Yet it does seem somewhat peculiar that a man with Jim's training and experience would not be sent to the field for four consecutive summers.

The Professor probably made one of the most rewarding trips of his career that summer when he was called upon to examine the land along the proposed route of the Grand Trunk Pacific Railway between Portage la Prairie and Edmonton. This trip through the so-called fertile belt was reminiscent of the kind of exploratory work he had undertaken for the Canadian government before joining the Survey; he was continually noting the vegetation cover and soil content, and he interviewed as many settlers as possible. There the similarity ended, for the once-unbroken wilderness of the 1870s was now a thriving grain-producing region. In fact, because of the destruction of the old trails, travel was so difficult that more than once Macoun's party was 'cornered by a field of wheat.'[6]

Although Macoun had once raised doubts about the region's capabilities in *Manitoba and the Great North-West*, any misgivings were conveniently forgotten and in his subsequent report to the government, he argued that his latest findings confirmed his earlier assessment.[7] He made similar claims before the Commons Committee on Agriculture and Colonization, before which he appeared for the third time in four years on 18 December 1906. After assuring the committee members that his statements could be taken 'not perhaps as Gospel truth but as scientific truth,'[8] he stated that the capabilities of the fertile belt had been seriously underestimated. In a visionary fit, he also announced that parts of the barren lands could be successfully brought under cultivation and that Canada as a northern country was destined to be a dominant society: 'After I am dead and gone and many of you also, this northern country will be a glorious country filled with happy people growing enormous quantities of wheat and other products. That is just as true as that the sun is shining today; there is not the slightest doubt about it. It is our wrong impression that is causing all our trouble. I am getting out of my wrong impressions, and I want you, gentlemen, if you have any, to get out of yours also.'[9] After further testimony along these lines, the committee recorded 'its appreciation of the valuable services professor Macoun has rendered to Canada in the past thirty years of his arduous and official services as a *practical* science officer of the Geological Survey of the Dominion.'[10]

The timing of this tribute could not have been better, for Macoun's 1906 assessment of the Grand Trunk Pacific route was the last survey of its kind that he would perform for the government. Throughout the museum debate during the 1890s, Macoun, among others, had justified the creation of a national museum on the grounds that the Survey's magnificent holdings needed safeguarding. Once the impressive stone structure began to take shape one mile south of the Parliament buildings, however, there was a change in thinking *vis-à-vis* the relationship between the museum and the Survey's collections. Although Macoun was proud of the herbarium, he was concerned about the number of outstanding gaps in the branch's zoological collections, particularly the absence of larger birds and animals. These shortcomings existed partly because too few men had been trying to collect too widely; there was only so much that the Macouns and Spreadborough could secure in the field each season. There was also little incentive to collect larger zoological specimens.[11] Not only were they difficult to preserve and transport back to Ottawa, but there was no space for them in the already overcrowded Sussex Street headquarters.

The anticipated move into the soon-to-be-completed Victoria Memorial Museum now promised to remove the display and storage restrictions and, in February 1906, Macoun advised Oliver that immediate action was necessary 'in order to make certain that we shall have exhibits worthy of the building when it is ready to be opened.'[12] What Macoun wanted was additional money to acquire some of the missing specimens – a request that he decided to take directly to Prime Minister Laurier a few days later. The Professor had been used to having access to government ministers during Macdonald's day and, although he disliked the Liberals, he saw no reason to discontinue the practice. 'If there is to be anything like a proper representation of our birds and animals in the new museum,' he wrote to Laurier following their private meeting, 'it will be necessary to have a special fund to be devoted to the purchase of desirable specimens and to pay for having them mounted once they have been secured.'[13] Laurier promised to act on the matter, and in June 1906 the government approved $3,750 for the purchase and preparation of museum specimens.[14] It was the first of several such grants for this purpose over the next few years.

The importance of improving the museum holdings was confirmed in the Mines Act of April 1907. Tabled by William Templeman, the current minister of inland revenue and minister-designate of the new department, the new legislation was significantly different from that which had been anticipated under Clifford Sifton. Whereas the former minister of the interior wanted to make the Survey more responsive to the mining interests of the Dominion and thereby threatened to alter dramatically the nature of the department, the government now moved to create a new Department of Mines with a Mines Branch and a Geological Survey Branch. The Mines Branch would fulfil those practical economic functions considered essential to Canada's mining and metallurgical industry, while the Survey Branch would continue to be responsible for its traditional, more scientific functions. Despite expectations to the contrary, then, the Survey emerged better off, in that it was relieved of duties that had been the source of criticism for several decades.[15] Indeed, it would appear that the Laurier government had learned a lesson from the Peace River report controversy.

The new Mines Act was undoubtedly a great source of relief to James Macoun. With Oliver no longer responsible for the Survey, there would be no future settling of accounts. Even the Professor could breathe a sigh of relief. Although the new legislation, unlike past acts, did not specifically state that the new Geological Survey Branch was to study the country's natural history, it did call upon it 'to collect, classify and arrange for exhibition in the Victoria Memorial Museum such specimens as are necessary to afford a complete and exact knowledge of the geology, mineralogy, palaeontology, ethnology, and fauna and flora of Canada.'[16]

Under Low's guidance, the Survey moved quickly to meet this responsibility, particularly in the area of natural history. The accumulation of biological specimens for the new museum, as evidenced by the activities of the 1907 field season, became one of the Survey's top priorities. While Spreadborough collected birds and small mammals along the west coast of Vancouver Island, Survey draftsman and explorer H.F. Tufts was temporarily engaged to secure specimens at the opposite end of the country in Nova Scotia. The Professor spent most of June and July in western Ontario and, despite a running battle with H.N. Topley over what particular trees should be photographed, managed to return with samples and pictures of more than forty trees of the region.[17] He then moved on to Percé, Quebec, at the extreme tip of the Gaspé Peninsula, and resumed collecting the seaweeds of the Gulf of St Lawrence, a project that he had started two years earlier. Geological parties, especially those working in remote or northern regions, were also instructed to secure specimens of larger game animals.[18] In trying to procure some of these mammals in Manitoba, one party ran afoul of the provincial game protection act, and the premier had to be prevailed upon to make special provision for the Survey.[19]

This tremendous activity in the field was matched by that at Survey headquarters. It was one thing to collect material for the new museum but quite another matter to make sure that it was properly prepared and ready to go on display once the building was completed. C.H. Young, an entomologist with the Experimental Farms Service, was consequently named assistant to the curator of the Museum on 14 April 1907 and turned over to the natural history branch.[20] Under Jim's direction, 'Bugs,' as he was affectionately known to his Survey colleagues, prepared a series of display cases illustrating the life histories of various butterflies and moths. He also made a complete inventory of the birds and mammal skins, noting those particular species that had yet to be secured.[21] Jim worked through old plant bundles from previous field seasons and tried to catch up on long-overdue exchanges. He found that the Survey herbarium had come to be regarded as a reference collection on Canadian flora and, as a consequence, that foreign botanists were increasingly requesting specimens of particular species or genera. 'There is no botanist in America or Europe who is not grateful to have the opportunity of studying our material,' he pointed out to Theo Holm, 'and the best men are working at it as fast as we can get it ready for them.'[22]

By the end of 1907, this concerted effort to collect and arrange natural history material for the Victoria Memorial Museum resulted in one of the most productive years of Macoun's twenty-five-year association with the Survey. Ironically, the very purpose of this activity – the approaching completion of the new museum – was not realized. The site that the Laurier government had taken more than two years to decide upon was found to be underlain with unstable blue clay. When the builder, George Goodwin, warned Public Works in August 1905 about the possible consequences, he was advised that the department had conducted tests and, going against geological knowledge, had concluded that the heavy stone structure would not settle to any extent.[23] Construction consequently went ahead, but in the late spring of 1906 the central tower came crashing down. Despite this unfortunate turn of events, the minister of public works assured the House of Commons that the building, complete with the newly dubbed 'Laurier Tower,' would be completed on schedule. The unstable ground, however, continued to make construction a nightmare, and the contractor spent almost as much time trying to stabilize the structure as he did building it. To make matters worse, the three hundred stone-cutters and masons that Goodwin had brought from Scotland for the job went on strike for a year.[24]

The failure of the new museum to be finished on schedule gave the natural history branch more time to complete and update its collections. This work would have to be carried out under new leadership, however, for Low had fallen seriously ill with cerebral meningitis shortly after the passage of the new Mines Act, and was henceforth effectively prevented from performing his duties. Geologist Reginald Walter Brock was consequently named acting director of the Geological Survey Branch on 1 December 1907. One year later, the government not only made this appointment permanent but also called upon Brock to serve as acting deputy minister of mines, a position that had been temporarily held by palaeontologist J.F. Whiteaves since May 1907.

Brock's tenure as Survey director represented a watershed in the institution's history.[25] A former professor of geology and petrology at Queen's University, he was generally regarded by the Canadian mining industry as one who had its interests at heart. He also shared the government's unbridled faith in the potential of the country and, because of his Liberal sympathies, got along so well with the Laurier administration that the Survey was funded to such an extent that the annual appropriation was often never completely spent. Brock's greatest strength, however, was his determination to end the wide-ranging, all-inclusive field work of the past, particularly during the Bell years, and to move towards greater specialization in the various areas of Survey endeavour. He saw a clear need to reorganize and consolidate the Survey's operations so that it was better suited to meet the needs of the twentieth century.[26]

One of Brock's major concerns was the retention of the Survey's responsibility for the Victoria Memorial Museum – a responsibility that he feared might be jeopardized if natural history was not given adequate attention.[27] Consequently, when Macoun approached him in January 1908 with a proposal to engage Spreadborough's services on Vancouver Island over the winter, Brock's response was that 'you [Spreadborough] may go to whatever place you think you can do the best work and that the Dept. will pay your ordinary expenses and the wages of a man.'[28] Spreadborough began work immediately, gathering small mammals and birds at various locations until he was joined by the Professor in late May. Attention then shifted to the sea, and for the next month the pair collected off Victoria before moving north to the Pacific Biological Station at Nanaimo. Macoun proposed to make Departure Bay his base of operations for the summer and had arranged several months before to use the boat, dredging equipment, and laboratory facilities of the research station. These plans had to be revised, however, when Jim Macoun, apparently angered at not being dispatched to the field that season as originally intended,[29] took a six-month leave of absence without pay, effective on 1 July; the Laurier government evidently still hesitated to give the assistant naturalist free rein. Bugs Young, who had been working under Jim's supervision in Ottawa, was consequently summoned to the coast to help with the dredging operations in Departure Bay. His presence at Nanaimo resulted in the mounting of several of the marine specimens on the spot. It also meant that the harvesting of the sea's hidden treasures could continue even after a reluctant Macoun had returned to Ottawa because of Jim's absence. When work was finally halted at the end of September, Young and Spreadborough shipped back to Ottawa 156 starfish, 195 crabs, 600 insects, 1,100 flowering plants, 400 mosses, 150 seaweeds, and an assortment of sponges, barnacles, and shells.[30]

Macoun was so happy with the results of the summer's dredging operations that he decided to spend another season on the west coast, working out of Barkley Sound on the ocean side of Vancouver Island. This kind of field activity was not something that was really needed at the time. The interests of the natural history branch and the new museum would have been better served if Macoun had organized his various collections from past seasons. It was not the time to be adding yet another field to his collecting list. Such a move also contradicted the activities of other natural scientists in similar institutions; most were not only confining their activities to a particular field or genus but also were studying the functional processes of organisms and not simply their external characteristics and distribution.[31] Macoun, however, continued to believe in the sanctity of traditional natural history and was not about to change his ways, particularly in light of his age. He was also anxious to tackle the field of marine biology in the knowledge that he might possibly find species new to science.

In the spring of 1909, then, the Professor, Spreadborough, and Young set up base at Ucluelet on the west side of Vancouver Island and settled down to a daily routine of collecting marine specimens by day and mounting them by night. The end result was another enormous haul. 'Between crabs and starfish and shells and seaweeds,' Macoun reported to Mrs Britton that December, 'I am so busy that I hardly know which way to turn.'[32] In fact, the success of the summer's dredging work was not realized until the Barkley Sound material was sorted and sent to American specialists for identification. W.H. Dall and P. Bartsch of the U.S. National Museum, for example, described sixteen species of shells new to science, naming one for each of the three collectors. Such recognition served to justify Macoun's field activities; as he told Dall, 'We are evidently experts in shell collecting.'[33]

While his father was wrapped up in his Vancouver Island dredgings, Jim remained confined to Survey headquarters, where he took advantage of the continuing delay in the completion of the new museum building to rearrange the herbarium. He also completed a revised edition of the long out-of-print *Catalogue of Canadian Birds*. The Professor had set to work on this new edition in early 1907 and part of it had been actually printed at that time. The manuscript was never completed, however, and in the summer of 1909, in an attempt to clear away all outstanding projects, Brock instructed that the work be rushed to publication as quickly as possible.[34] Jim dropped everything else and managed to get a new one-volume edition ready by late October.

The bringing together of a mass of bird observations over such a wide range of territory in a single volume was no small accomplishment. And because of Jim's close attachment to William Spreadborough, special attention was drawn to the field assistant's ornithological field work over the past two decades. 'It detracts nothing,' observed the preface to the new edition, 'from the importance of other notes published for the first time in this catalogue to say its chief value is to be found in the matter credited to Mr. Spreadborough. His notes, revised by us, cover nearly the whole Dominion from Labrador and Hudson Bay to Vancouver Island and north to the Peace River.'[35] In his haste to complete the new edition, however, Jim had failed to correct several errors in the original three-part catalogue and to include some of the more recent field data. The new bird catalogue consequently came under immediate criticism from those knowledgable in the field: 'It is a peach,' lamented Percy Taverner to fellow ornithologist J.H. Fleming upon finding that some of his observations were still credited to Spreadborough. 'As far as I can see this new edition is not a whit better if as good as the old one.'[36] Jim, for his part, blamed the catalogue's shortcomings on his heavy work-load and promised to publish 'and [sic] addendum that will include not only additions but corrections' so that 'by this time next year we will have Canadian ornithology brought right up to date.'[37] The truth of the matter, though, is that neither he nor his father had the expertise to produce the kind of professional publication that would satisfy ornithologists. The original *Catalogue of Canadian Birds*, despite its obvious flaws, had been warmly received because it was recognized as one of the first major attempts at such a work – a necessary starting point. Simply to reissue a new, enlarged edition with essentially the same mistakes, however, was not acceptable and made a mockery of the continuing advances in the field. The Macouns apparently realized this, for after the reception of the new bird catalogue, they abandoned any plans to publish manuscripts in other areas and concentrated on general collecting and botany. Although they continued to collect fish whenever possible, there was no further mention of a catalogue on the topic. The proposed catalogue of Canadian mammals would also never be published. In manuscript form by this date, the work was still several years away from completion and lies today, forgotten, in a work-room in the vertebrate zoology division of the National Museum of Natural Sciences.

With the Victoria Memorial Museum expected to be occupied some time in 1910, there was a concerted push by the natural history branch that year to collect as much and as widely as possible. Activity was therefore spread over three different fronts. Given the success of the past two field seasons' dredging operations on Vancouver Island, a similar campaign was carried out in Nova Scotia: while Bugs Young dredged at strategic points along the coast, Macoun made a reconnaissance survey of the flora of the peninsula. Spreadborough, in the meantime, was busy on the west coast. Over the 1909–10 winter, he had remained behind at Ucluelet with the intention of collecting sea birds. By January, however, his fingers had become so badly poisoned by the arsenic that he used in the preparation of specimen skins that he had to forsake this work in favour of collecting marine specimens with the assistance of the local fishermen. Every now and again, he would visit their boats to examine their catch. Once he even brought along a bottle of booze in his pocket, offering them a drink 'to keep their eyes open for any rare things that may get tangled in the nets.'[38] When his fingers finally recovered, he was sent north for the summer to Skidegate in the Queen Charlotte Islands, a region that had last been examined by George Mercer Dawson in 1878.

Even Jim was active in the field that summer. Following the completion of the revised edition of the bird catalogue, he started to work through some of the field bundles that had not been opened since the day they had arrived at the office. These old collections, according to Jim, were 'chiefly leftovers when my father and I would leave for the field before we finished with the collections of the previous year, and in fact these leftovers in some cases constituted the bulk of the collection.'[39] But before he got very far, he was advised of the government's decision to send him to the northwest coast of Hudson Bay to make a collection of the local flora and fauna for the new museum. Jim was probably given the nod for this trip because of Spreadborough's poisoned fingers and, despite his hard feelings about past failures to send him to the field, he was quite excited at the prospect of getting away to the north again. Determined to show how wrong the Laurier government had been in confining his activities to Ottawa since 1906, he sought Dr Merriam's advice as to what specific birds and mammals should be secured. 'If I go there,' he wrote, 'I want to work on the lines that are of the greatest scientific value.'[40]

Jim reached Churchill on 25 July 1910 and spent nearly a month working up the natural history of the region. He then boarded the Newfoundland-based wooden schooner *Jeanie*, which had been chartered by the local North-West Mounted Police detachment to transport police supplies and portable houses to various points along the west coast of Hudson Bay. This arrangement promised to suit Jim's needs perfectly, for it would enable him to collect over a wide range of territory with relative ease. The *Jeanie*, however, did not inspire confidence. She was lucky to have made it as far as Churchill and was already sporting a gaping hole in her patchwork sail. The crew was little better. 'I would not like to be stranded on the beach up north with a crew of such fellows,' Jim would remark later.[41]

Departing from Churchill on 24 August, Jim spent two days collecting at both Cape Eskimo and Rankin Inlet, while portable houses were erected. The planned visit to Chesterfield Inlet had to be abandoned because of rough weather, and the schooner made for Cape Fullerton where she was nearly wrecked. Despite this close call, the *Jeanie* limped northward to her final stop, Wager Inlet, but not before taking an unexpected detour via Southampton Island because of a sluggish compass. While the cargo was being put ashore on the afternoon of 9 September, a snowstorm with gale force winds came up, broke both anchor chains, and drove the ship aground late that night. When daylight broke, everyone on board walked safely ashore. Jim, for one, did not regret the loss of the schooner and was just happy to unload all his Churchill specimens that had been placed on board for transport back south to Newfoundland.

Jim had been shipwrecked before, almost twenty years earlier. In 1892, during his first trip to the Pribilof seal rookeries, he had nearly drowned when his ship struck a rock off the Queen Charlotte Islands.[42] He now drew on this experience and, together with one of the NWMP constables on board, took charge. A gasoline launch and whaling boat that had been used to ferry cargo ashore were made ready for the trip south along the coast. Because of space limitations, baggage and any extra clothing had to be left behind at Wager. Jim refused, however, to part with his collections, and 'it was only by intimidation that I was allowed to take my specimens with me.'[43] Setting off on 16 September, the two small craft took three days to reach Cape Fullerton, where they met a whaling schooner that took them down to Churchill. Since it was late in the season, Jim could not get back home by water that fall and consequently did some further collecting in the area until overland travel was possible. He finally started south by dog-team on 5 December, bringing with him nearly two hundred pounds of natural history specimens. He reached Gimli, just north of Winnipeg, six weeks later and twenty-three pounds lighter, after travelling 1,100 miles in minus-forty-degree weather. After he telegraphed his whereabouts to a relieved father and family, he made for Winnipeg and eventually reached Ottawa, with specimens in hand, on 18 January 1911.[44]

By the time Jim arrived back in Ottawa, the Geological Survey had been transferring material into the Victoria Memorial Museum for two months. This move was premature, in that the government had not yet assumed possession of the structure because of construction problems. The resurrected 'Laurier Tower' had begun to slump again, creating an ever-widening crack in the outside wall and the separation of the interior floors and walls. Labourers, meanwhile, refused to work in the basement because of the possible danger of flying stone chips as the building slowly settled into the ooze.[45] Brock feared, however, that if occupation of the structure were delayed, the Survey might lose control of the museum and its operations.

Several months earlier, Clifford Sifton had sent Prime Minister Laurier a draft act in which the 'National Museum' was placed under the control of the Commission on Conservation, a non-partisan body that had just been created by Parliament to make recommendations for the more efficient development and conservation of Canada's natural resources.[46] When asked to comment, Minister of Mines William Templeman objected in the strongest possible terms. He saw the proposal as nothing less than a bold attempt by Sifton, as commission chairman, to secure some administrative and executive power for the otherwise purely advisory body at the expense of the Department of Mines. 'No good purpose would be served,' he warned the prime minister. 'The museum will accomplish everything aimed at ... by leaving its control and management where it is.'[47] Nothing therefore came of the Sifton challenge, but the Geological Survey was taking no chances. At the conclusion of the 1910 field season, the transfer of men and collections to the new facilities began in earnest.

The timing of the move could not have been more opportune. In early December 1910 Laurier received first a petition and then a delegation, which included Macoun, calling for the museum to be placed under the control of a special government commission along the lines of the British Museum and the United States National Museum. 'We feel very strongly that now ... is the time to establish a National Museum in the Capital, the Washington of the North,' the petition read, 'that will educate Canadian citizens who will study it and inform all strangers from other lands, who may visit it, as to the wealth of heritage which a kind Providence has dowered upon us, and which Young Canada, a Nation in The Making, is working to develop.'[48] The proposal was too late, however, for the Survey's early move into the new building had effectively settled the issue. Besides, the Laurier government probably wanted to avoid any controversy; the museum was already three years behind schedule and $300,000 over budget.[49]

In taking part in the delegation to the prime minister's office, Macoun probably believed that the natural history branch could best prosper under a museum commission. He had been dealing with the British Museum and the u.s. National Museum for years and had always been envious of the working environment. Yet the Professor did not have to worry about the Survey's administration of the new museum. For one thing, the collection of natural history material had been an official duty of the Survey since 1877 – a duty that was confirmed in each successive act. The ability of the Survey to administer the Victoria Memorial Museum, moreover, had been seriously questioned over the past few years. If Survey control was to be maintained, Brock realized, the organization and activities of the natural history branch – in fact, of all departments – would have to be augmented.[50] The first move in this direction had already been made in 1910. With the death of J.F. Whiteaves the previous year, Brock used the opportunity to end the unnatural grouping of palaeontology and zoology by creating a new natural history division. Under this arrangement, seventy-nine-year-old Macoun, as the new division chief, now had complete authority over the natural history work of the Survey, directing a four-person staff composed of his son, taxidermist Herring, invertebrate zoologist Young, and a full-time stenographer.

Further changes were made following the natural history division's move into the new building in late January 1911 − a move that was hurried by an outbreak of typhoid fever in Ottawa and the subsequent takeover of the old Survey museum as a temporary hospital.[51] Since assuming Low's duties as director, Brock had regularly spoken of the need to secure additional qualified personnel. In the summary report for 1910, for example, he had vowed 'to strengthen the staff, particularly in divisions that are relatively weakest' and to make the Victoria Memorial Museum a 'complete natural history museum.'[52] Brock took the first step in fulfilling this pledge when he hired Percy Taverner as assistant naturalist and curator in charge of vertebrates effective 1 May 1911. The appointment of the Canadian-born ornithologist, who had practised as an architect for ten years in Chicago and then Detroit, was not surprising. Some five years earlier, when Macoun had discussed the need for a museum curator with J.H. Fleming, the wealthy Toronto ornithologist had put forward Taverner's name. The matter was dropped, however, when it was learned that Low was more interested in securing a full-time taxidermist to replace Samuel Herring.[53] Then, as the museum neared completion in the spring of 1910, Fleming, along with fellow birder W.E. Saunders of London, Ontario, tried once again to secure Taverner's appointment. 'Besides being a keen observer, an extra good taxidermist,' Saunders informed Macoun, 'he has a very artistic temperament ... I am sure it would be exceedingly difficult for you to find another man so well equipped to assist you.'[54] Well aware of Taverner's credentials and badly in need of help in zoology, the Professor was won over. So too was Brock. The curious thing about Taverner's appointment, however, was that no one at the Survey had discussed with him the exact nature of his duties. When Taverner subsequently sought clarification, Jim told him that he would be responsible for organizing the Survey's specimen collection and getting displays ready for the museum.

Beyond that, Taverner was to be his own boss: 'Your work, as I understand it will be practically independent of everything but general instructions.'[55] The reason for this open-ended arrangement was soon apparent to Taverner. After only two weeks on the job, he wrote to Fleming that Macoun 'is a great field man but of very little use as a museum head and I think realizes it and wants someone who is perhaps to cover up this weakness in him and perhaps in Jim who is scheduled to take his place.'[56]

Around the same time that Taverner was appointed, the Macouns decided to approach Brock about the possible hiring of Dr Edward Greene of the United States National Museum for the vacant position of Dominion botanist. That the natural history division should take on another botanist would seem rather odd. Both men had considerable field experience in the area and any help that they needed was freely given by their foreign colleagues. The Professor, however, was becoming increasingly conscious that he was one of the last of his generation of botanists and was worried about the future. Jim, on the other hand, had serious doubts that he would ever succeed his father; his Peace River pamphlet had sealed his fate.[57] The most important reason for trying to secure the services of someone like Dr Greene, however, was the need to have a recognized botanical authority at the new museum who could name and describe plants new to science.[58] As field collectors, neither father nor son could perform this kind of specialized work, and they decided to take advantage of the move into the new museum to resolve this problem. Otherwise, despite its modern facilities, the natural history division would continue to rely upon outside experts for assistance. In a sense, the decision to try to hire another botanist was a tacit confession of the Macouns' limitations. And it was something that they had been thinking about while the museum was under construction. In 1906, Jim had secretly written to M.L. Fernald of the Gray Herbarium, Harvard University, inquiring whether he was interested in the position of dominion botanist.[59] Now, they pinned their hopes on Dr Greene. Brock believed, however, that other areas of the museum's natural history work required bolstering and would only agree to the appointment of a botanical specialist at some future date.[60]

The natural history division's first year of operation in the Victoria Memorial Museum was not marked by the undertaking of any significant new ventures commensurate with the larger, more spacious facilities. Although the move into the new building had been actively prepared for for half a dozen years, much remained to be done in terms of assessing and reorganizing the natural history material accumulated since 1843 – much of it by the Macouns over the past thirty years. Such work had been quite impossible in the cramped quarters of the old Sussex Street building. The natural history division did not know exactly what it had until the material was rounded up and transferred. 'When we moved from the old Museum,' Jim told Theo Holm, 'the hundreds of bundles of unworked over specimens that were scattered about everywhere were all got together and it was enough to frighten one only to look at them.'[61] The first few years in the Victoria Memorial Museum were therefore largely devoted to the organization of the division's operations and the preparation of displays. Except for Macoun's activities, any field work was limited and generally concerned with the completion of the museum collections.

Following the move into the museum, the Professor completed a draft manuscript on the flora of the Maritime provinces in consultation with local botanists Dr G.U. Hay of Saint John, Dr A.H. Mackay of Halifax, and Lawrence Watson of Charlottetown.[62] He then turned his attention to a long-planned flora of the Ottawa region and spent the summer roaming within a thirty-mile radius of the capital. On at least two occasions, Macoun took Taverner to the field with him on short day trips to see the ornithologist in action. It was Taverner, however, who returned home impressed. 'It is most astonishing to see Prof. Macoun in the field,' he wrote to Fleming. 'Works through swamp, thicket and meadow with all the energy of a young man. He is a wonder.'[63]

Now that Taverner had assumed responsibility for the vertebrates, Macoun had given up any thought of covering all fields of natural history and had set his sights on 'perfecting our knowledge of the flora of our country.'[64] He was more than ever concerned, nonetheless, about the growing number of botanists who specialized in a particular area and spent more time studying plants in the laboratory than in their natural habitat. 'Plants must be collected and studied in the field,' he railed against specialism to a fellow botanist, 'before descriptions can be absolutely accurate.'[65] Jim also worked at botany. By rough estimate, it appeared that it would take at least two full years to work through all the material that required his attention.[66] He had to process the large collection of plants that he had made on the west coast of Hudson Bay, as well as all the old, unopened bundles from previous field seasons. Yet before Jim got too far into this work, he and W.A. Found of the Department of Fisheries were named as expert witnesses to the British delegation at the North Pacific Fur Seal Convention and, on 10 May 1911, accompanied Joseph Pope, under-secretary of state for external affairs, to Washington, DC. For the next two months, Jim drew upon his firsthand knowledge of the Pribilof fur seal rookeries to assist Pope and the British ambassador to Washington, James Bryce, in reaching an agreement with the American, Russian, and Japanese representatives over how the herds were to be conserved and the annual catch divided. The bargaining was tense and at times the conference threatened to break down, but finally an agreement was reached on 7 July 1911. According to the British negotiators, Bryce and Pope, it was a splendid bargain.[67] Jim, for his part, was indifferent to the outcome, for his twenty-one-year-old daughter, Helen, had died during the early stages of the negotiations. 'The Seal Conference was just a little episode in my varied life,' he wrote to Dr Greene that October, 'and I have forgotten about the little beasts.'[68] In fact, he had no sooner returned to Ottawa than he took to the field with his father to try to forget about his loss. The Canadian government, however, did not overlook his contribution and paid him a $500 bonus. The British were also thankful. On the recommendation of Bryce, Jim was awarded the Companion of St Michael and St George (CMG) in 1912. Ironically, even when things worked out for the assistant naturalist, his uncanny knack for calamity tripped him up once again. On the day that he was to receive the award, he called into a local bar for a bracer. When one of his socialist friends learned that he had accepted the award, he apparently yelled 'traitor' and punched Jim. That night, when Jim arrived at the Government House dinner party for the formal bestowal of the honour from the Duke of Connaught, he sported a black eye.[69]

The rest of the natural history division's staff devoted most of their energies to the Victoria Memorial Museum collections. Bugs Young was given the long, arduous task of getting the Survey's invertebrate holdings ready for exhibition and, except for a ten-week trip to New Brunswick in the late summer, spent the entire year on this project. Spreadborough, on the other hand, remained at Victoria for the summer, filling any remaining gaps in the Survey's island collection. The bird and mammal specimens that he sent to Ottawa were now handled by Taverner, who had been saddled with the major task of reorganizing the vertebrate material. The new assistant naturalist warmed to the challenge, however, and within twelve days of starting work at the museum, had a detailed, nine-page evaluation of the current state of the vertebrate collection on Brock's desk.

Taverner's report reflected the problems that the natural history division had faced in trying to undertake a survey of the biological life of the Dominion with limited manpower and financial resources. In many ways, it was an indictment of the zoological work that Macoun, his son, and Spreadborough had performed under Survey auspices. The report found that although there was a sizeable collection of mounted material, much of it was old work done by obsolete methods and suggested that it should be gradually weeded out of the collection. The study specimens were not much better, but because many species would be difficult to replace, the report recommended that they be 'relaxed, thoroughly cleaned ... and made into skins again.' Taverner was equally critical of the labelling and cataloguing of specimens. Not only was there no apparent scheme of keeping track of what was in the collection, but the specimens themselves were designated 'by hurried field labels some made with pencils on scraps of paper and with localities very vague and unsatisfactory ... almost undecipherable from grease and age.' In essence, then, the existing vertebrate collection needed a complete overhaul, and one of the first steps in this direction would be the acquisition of a properly trained preparator. The Survey's long-time taxidermist, Samuel Herring, simply would not do. 'The least said about our present taxidermist the better,' Taverner advised, echoing Macoun's sentiments of some twenty years earlier. 'He is an old man versed in the ways of fifty years ago ... If he shows any willingness he can be a useful man, otherwise he takes up valuable room.' The report also warned that 'for the time being and indeed for some years to come we will have just about all we can do collecting and studying materials from our country.'[70] Although Macoun regularly boasted about the size of his field collections, the Survey's faunal holdings were actually far from representative. The best Canadian bird collection, for example, was held by Taverner's Toronto mentor, J.H. Fleming.[71] Taverner consequently suggested that Survey parties not only continue the traditional policy of gathering biological specimens but that rarer items be purchased from private collections. Finally, the report recommended that a survey of other similar museums was necessary before any policy on the nature and development of the displays could be finalized. 'I am interested in making this institute take its place among the great institutions of the world,' he concluded. 'My scientific future is wrapped up in it ... and incidentaly [sic] of all who have connection with it.'[72]

Despite its many criticisms, Macoun welcomed Taverner's report and advised that it be sent on to the director;[73] it was one of the few times when the Survey's natural history work took precedence over his ego. The Professor, however, had given up his zoological studies and was just happy to have some of the problems associated with past work in this area dealt with in an effective, constructive manner. The report, in particular the suggestion to learn from the successes and failures of other institutes, also struck a responsive chord with Brock. Although he and the minister of mines had just returned from an inspection tour of American museums the previous summer, it was decided to send the new assistant naturalist south, since he was responsible for the arrangement of the vertebrate collections. Besides, money was no longer an obstacle. Whereas previous directors had laboured under severe financial restrictions, Brock was being granted more money than he could spend; in 1911, for example, the unexpended balance was $120,594.81.[74] Taverner consequently left Ottawa on 15 June and visited some of the larger museums in Boston, New York, Philadelphia, Washington, and Chicago. Upon his return six weeks later, he issued another report, which expanded and enlarged upon his earlier remarks. Apart from reiterating the need for a good taxidermist, he noted that the addition of a herpetologist and mammalogist was 'quite necessary if we are to do work that will measure up to the standard set by the serious museums on the other side.'[75] He also called for the creation of special study and storage facilities in the new museum, as well as the development of a proper system for cataloguing the collections. He strongly believed that there would be little constructive exhibition work of any kind until these changes were forthcoming.

Having placed these recommendations before Brock to mull over, Taverner then turned to the matter of getting the vertebrate material into some reasonable order. It was not a simple matter, however, of sitting down and assessing the collections. Before actual cataloguing could begin, the history of each particular specimen and accompanying field data had to be verified. This preliminary work proved to be a tedious operation, 'involving the most careful deciphering of obliterated labels, searching of maps for little-known localities, and research among the various summary reports and old records scattered through many registers and manuscript lists for years back.'[76] In many cases, Macoun's memory was the only source of information.[77] Taverner worked steadily at this job for the remainder of the year, but even with the assistance of artist Frank Hennessey, he was able to do little more than relabel most of the birds. Still, it was an important beginning to a new era in the Survey's natural history work.

Macoun was quite pleased with the events of 1911. The natural history division had finally occupied the Victoria Memorial Museum with its long-awaited exhibition and storage facilities. The vertebrate collection had been placed in the hands of a competent naturalist who had commenced the proper cataloguing of the specimens and had mapped out the future needs of the division. And finally, the party that had appointed Macoun in 1882 was back in office after winning the September 1911 general election. The Professor, for that matter, could look back over the past thirty years with a great deal of pride and satisfaction. The Survey's natural history work had undergone a remarkable transformation since his first days as Dominion botanist. Macoun, with the assistance of his son and Spreadborough, had seen to it that the Survey's biological duties were upheld and not simply given passing attention. This work was not without its problems and shortcomings, but on the whole the situation could have been much worse without their dedication and drive. Much of what they had accomplished would serve as the basis for the National Museum of Canada.

Macoun chose this moment to bow out. Although the future of the natural history division, renamed the biological division in 1912, seemed bright, he did not see himself as part of it. Not only had many of his Survey colleagues retired or died, but it was also quite evident from Taverner's activities that the Professor's work was rapidly passing to a younger generation. His wife, Ellen, had become ill in May 1911 and could no longer be left on her own. The Professor therefore decided in February 1912 to semi-retire to Vancouver Island and live with his eldest daughter, Clara, at Sidney, British Columbia. 'I am looking forward to doing good work for years to come on the Pacific Coast,' he advised a friend, 'where I fully intend to live out my later days.'[78] In the excitement of getting ready to move to the coast, however, he suffered a paralytic stroke on 6 March that incapacitated him for six weeks. On 26 April 1912, in the company of his wife, his son James, and James's family, the Professor bade farewell to Ottawa and his work over the past three decades and boarded a westbound train to return to the simple joys he had known as a private collector.

The End of an Era

John Macoun's departure for Vancouver Island did not end his association with the Geological Survey of Canada. For the past three decades, the Survey's natural history work had revolved around the Professor, and the man and his work were not easily separated. Despite the fact, then, that he was living some 2,500 miles away, he continued to hold down the position of chief of the biological division until he was succeeded by his son in 1917. The rest of the division, meanwhile, when not busy sorting, assessing, and cataloguing material from past field seasons, remained preoccupied with trying to make the Victoria Memorial Museum's holdings as representative as possible – elementary work more in keeping with the nineteenth century. Any expectation that they would be engaged in the kind of specialized biological research that was associated with the new facilities vanished with the outbreak of the First World War in 1914; their work continued to be inhibited by the lack of manpower and financial resources. The 1910s were therefore years of delay, frustration, and disappointment for the biological division – a difficult decade capped by the death of both John and James Macoun in 1920.

By the time he reached Sidney on 29 April 1912, Macoun had made a remarkable recovery: the only lingering side-effect of the stroke was a partial lameness in his right side that forced him to write left-handed and to walk with the aid of a cane.[1] It was apparent from the outset that the Professor had no intention of settling down to a quiet life with his daughter's family. Arrangements were immediately made for the construction of an addition to the Wheeler home, which Macoun subsequently named 'Ninety-Eight' after his long-time Ottawa residence, 98 James Street. A small cottage across the street was also secured to serve as his laboratory and herbarium.[2] Although he remained deeply interested in what happened at the new museum, he was tired of fighting the same old battle and preferred to leave matters in his son's and Taverner's hands. More than anything else, Macoun simply wanted to spend his remaining years wandering at will in the field. When some of his old Survey colleagues who were visiting Victoria later that summer called on him, 'They expected to find an invalid confined to his chair if not his bed. To their surprise he was out in the woods when they arrived and it was some time before he returned – with trousers torn in the bush and without a cap or hat but happy and with a basket of plants.'[3]

While Macoun was getting established on the west coast with the assistance of his son, the other members of the biological division pushed ahead with the seemingly endless task of properly documenting and accessioning the museum collection. Although Bugs Young continued to sort through the invertebrate material for the second straight year, he still had considerable work ahead of him. Like the Professor, the late Dr Whiteaves had left the Survey's invertebrates 'in a state resembling chaos as far as data goes,' and poor Young was 'struggling desperately to work out the localities.'[4] Taverner fared little better. By the time the Macouns had left for Vancouver Island, he had managed to complete the cataloguing of the Survey's bird collection – a collection whose value was greatly enhanced over the next few years with the purchase of a number of private collections, including his own.[5] He then turned his efforts to the mammal collection but, like the Macouns in the past, found that his time was constantly being consumed by other demands. In order to ensure that future field collections would meet the new standards he had established over the past couple of years, he prepared a manual for the proper collection of zoological specimens, as well as a new field notebook for recording natural history data.[6] He was also somewhat embarrassed by the division's failure to have anything on display after more than a year in the new building and, while awaiting delivery of the new display cases, assembled a few temporary exhibits 'to make a little show for the public.'[7]

Taverner also became involved in the formulation of general museum policy when, in Jim's absence, he was named the biological division's representative on a newly struck Survey museum committee. As a relative newcomer to the Survey, he believed he was not the best person to decide such things as the organization of staff and the assignment of exhibition space. He dearly wished that Jim was present in Ottawa to speak for the museum's natural history endeavours.[8] Taverner was better off on his own, however, for Jim regarded the formation of a museum committee as a complete waste of time. He believed that the biological division should simply be given a free hand to do as it saw fit. 'Poor Brock thinks that if he gave us a chance we would knife him,' he privately advised Taverner from Sidney. 'He doesn't seem to realize that after wasting our lives plugging we would all hurrah for anyone who would give us a chance to do some real work.'[9]

During the museum committee deliberations, Taverner kept the Professor abreast of its recommendations and sought his advice on proposed exhibits, particularly those pertaining to botany. He also used his position on the committee executive to push for the hiring of a number of biological specialists. He was depressed about the valuable time he was spending on routine sorting and cataloguing at the expense of research and field work, and in his summary report for 1912 argued that staff additions were absolutely necessary 'if our museum is to take its place among sister institutions.' Until such help was forthcoming, there would be 'no time for original or other work.'[10] In the end, the biological division secured not one but two new staff members to fill major gaps in its ranks. On 1 May 1913, Clyde Patch, a former preparator with the American Museum of Natural History in New York, was named chief taxidermist to the Victoria Memorial Museum. His duties were subsequently expanded in 1918 to include herpetology.[11] The second appointment was completely unexpected. Fresh from his successful 1908–12 expedition to the central arctic, the Canadian ethnologist-turned-explorer Vilhjálmur Stefansson proposed a major reconnaissance survey of the Beaufort Sea region. Stirred by the prospect of the exploration of new territory, the Borden government decided to sponsor the project, and in February 1913 the Canadian Arctic Expedition came into being. Dr Brock used the opportunity not only to send a few Survey officers along but also to hire some of the men that Stefansson had himself lined up for the expedition.[12] In particular, Dr Rudolph Martin Anderson, an American-born mammalogist who had taken part in the earlier Stefansson expedition, was appointed assistant zoologist, effective 1 June 1913.[13] Unfortunately, Anderson's subsequent departure as leader of the Canadian Arctic Expedition's southern party meant that the biological division would not receive the benefit of his services until his return some three years later. Still, it could take comfort in the thought that it would soon be the recipient of natural history specimens from Canada's western Arctic region.

Jim, in the meantime, had returned from the west coast with his own plans to secure additional personnel. Never having lost sight of the need to have someone associated with the museum who could name and describe plants, he decided to take advantage of his father's semi-retirement to Vancouver Island to ask Dr Brock for the appointment of a botanical specialist. In a March 1913 memorandum on the current state of the botanical branch, he reported that even after another winter of working up past collections, there were still many specimen bundles dating as far back as 1908 that had yet to be identified and catalogued. In fact, Jim's memo created the distinct impression that they would probably not be examined in the immediate future unless he abandoned his field work.[14] Dr Brock had other thoughts on how the problems of the botanical branch could be resolved. Since many of the unprocessed collections had been made by the Professor, he reasoned that the old man should help clear up the backlog before proceeding with his Vancouver Island work. The director consequently asked Macoun to return to Ottawa 'to help your son straighten things out here and to get the results of past work into such form that it will not be lost but can be utilized by everyone.'[15] The Professor, however, had no desire to leave the island to work up past field collections and, without consulting Brock, had written to the ministers of labour and the militia, two old Conservative friends, 'telling them what he would like done' in recognition of his thirty years' service with the Geological Survey.[16] It was subsequently established by an order-in-council dated 9 June 1913, that Macoun would retain his position as division chief but that his activities would be restricted to the west coast and that he would be paid only while engaged on official field work.[17]

This milking of the political system probably upset Brock, since he was trying to build a professional Survey staff. As in 1887, it also had a cost for Jim. When in November 1913, the assistant naturalist sought the appointment of a preparator-collector who could handle the routine herbarium drudgery 'that has been done by me and nearly all of it after or before regular office hours,'[18] nothing came of the request. Jim's disappointment with the direction things were taking at the museum was captured by his father shortly thereafter. 'Jim seems to think there is a new atmosphere at the Survey,' he wrote to Taverner. 'We used to be a band of brothers on the Survey but Brock destroyed the fellowship.'[19]

While Jim tangled with Brock in Ottawa, his father was busy compiling an impressive collecting record on Vancouver Island. His accomplishments by the end of 1913 were nothing short of astounding. Since coming to Sidney, he had collected 247 species (937 specimens) of fungi, 128 species (605 specimens) of lichen, 31 species (118 specimens) of liverworts, 700 species (3,000 specimens) of flowering plants, and 195 of the known 264 species of island seaweed. He had also found an extremely large number of mosses but had not yet arranged them.[20] As in the case when he had first started taking a serious interest in botany, all his collections had been named by specialists and then mounted and placed in his little herbarium. And as in his earlier years, he continued to find new forms. In addition to twenty species of flowering plants new to the island flora, he had found eight species of fungi new to science.[21] As far as Macoun was concerned, however, he was just getting rolling, and he told Brock in the spring of 1914, 'I mean to commence a more extensive system of field work.'[22]

The other members of the biological division were more active in the field that summer as well. For the first time since joining the Survey staff, Taverner, with Young as his assistant, was able to slip away from the Survey for more than a few weeks to conduct a fairly extensive survey of the Chaleur Bay region of Quebec. The notes from such outings were to be used in a new Canadian bird book that Taverner was preparing. Jim roamed farther afield. He collected on the northern end of Vancouver Island and then continued north to the Pribilof Islands as the Canadian representative on an international commission appointed to study and report upon the condition of the North Pacific seal rookeries. Such field activity was now possible because the business of getting settled in the new building was nearly complete. In the case of zoology, not only had most of the pre-1912 museum collection been catalogued and arranged but the office and preparatory department had been organized so that it was better equipped to deal with all incoming collections. Even botany had made significant progress: although there were still several thousand specimens from previous field seasons to examine, the bulk of the herbarium material had been put into new wrappers, labelled, and then placed in new cabinets. 'For the first time in thirty years,' Jim proudly reported, 'it may be said that all the botanical material in the herbarium has been put into the condition that it is readily available for study and reference for anyone.'[23] The number of mounted herbarium sheets was almost double that of only five years earlier.

Once the collections had been properly organized, the biological division came face to face with the problem that the Victoria Memorial Museum was the responsibility of the Geological Survey and not a separate entity unto itself. The interests of the museum consequently sometimes suffered. When, for example, the new building had been occupied in 1910–11, the rapidly growing drafting and topographical staff had moved in as well. The presence of these officers, whose duties had nothing to do with the museum, inevitably resulted in overcrowding and the misuse of space. The offices, laboratories, and workrooms were so congested that they were literally overflowing with material, while valuable exhibition space was being used 'as a freight shed and as a storage room for tents and other field equipment as well as a lot of other materials which for lack of a more suitable definition may be called rubbish.'[24] It seemed as if the problems of the old Survey building on Sussex Street had returned to haunt the new museum.

The museum committee had tried unsuccessfully to draw attention to this situation in July 1912. A second attempt was made in October 1914, after Dr Brock had resigned the Survey directorship in favour of a post at the University of British Columbia and been replaced by another geologist, R.G. McConnell. In a carefully argued memorandum, the museum staff explained in considerable detail how the success of any museum, no matter how impressive a structure, was largely dependent on its preparation department and storage facilities. 'No man exercising sound judgment will go to the expense of erecting a building on definite and unmistakable lines,' it observed, 'and then permit its normal intent to be frustrated by the presence within it of activities foreign to its purpose.'[25] The memorandum then outlined how the presence of the drafting and topographical staff was adversely affecting the activities of the museum. It was rapidly reaching the point where future acquisitions might have to be turned down. Something had to be done, the memorandum concluded, 'if failure is to be avoided.'

There was little that McConnell could have done to ease the crowded conditions at the Victoria Memorial Museum, even if he had wanted to. With the outbreak of the First World War in August 1914, the importance of museum work had greatly diminished relative to the Survey's other functions – especially locating and developing of Canada's strategic resources.[26] Given the demands of the war, moreover, the situation called for consolidation and stability, not the displacement of staff for the sake of putting some stuffed birds or animals on display. The biological division and the museum in general simply had to make do with what they had. The Professor, for his part, remained ever confident. 'I am very glad to hear how well you are getting along,' he confided to Taverner, 'and pleased to know that my work was not in vain. I am sure that you will have (as Young would say) a big show yet.'[27] It certainly appeared that way, for in 1915 the biological division acquired a record number of zoological specimens.[28] All the museum needed, in the words of the 1914 memorandum, was 'a freedom of action which it had not yet experienced.'[29]

The situation, however, only worsened. On the night of 3–4 February 1916, the centre block of the Parliament buildings was destroyed by fire. Since Parliament was in session, it was immediately decided to establish temporary Commons and Senate chambers in the lecture and west exhibition (fossils) halls of the Victoria Memorial Museum. The museum staff consequently had to vacate the building, while the collections were boxed and placed in storage.[30] As if this upheaval were not enough, the new museum building itself was dramatically altered. The infamous 'Laurier Tower,' which had begun to separate from the rest of the building, was removed before it could collapse again. The great hopes for the Victoria Memorial Museum thus continued to be foiled. If anything, though, the squat, fortress-like appearance of the structure was in keeping with its new occupants – a wartime government.

The biological division had faced similar crises in the past and survived; this current crisis was no different. Jim was on good terms with McConnell – the two men had known each other for more than thirty years – and convinced him that they would need several months to pack up material.[31] Unlike other divisions that had to vacate the building on short notice, then, biology was allowed to remain in the Victoria Museum until January 1918, when it finally moved into its new quarters in the Lowe-Martin Building on Nepean Street. Ironically, these temporary facilities were considerably larger than those the division had previously occupied and served to underline the poor conditions at the new museum.

Jim's fortunes during these years were also mixed. In late 1916, he was appointed to the five-man Arctic Biological Committee that had been created to oversee the working up of the field results and the publication of the reports of the Canadian Arctic Expedition. The following spring, McConnell recognized the fact that the assistant naturalist was running the biological division in his father's absence and appointed him acting head; the position was made permanent 28 February 1918.[32] Jim quickly learned, however, that his increased status did not bring with it an escape from the task of working over the outstanding collections from previous years. As he explained to American botanist M.L. Fernald:

I spend at least two or three hours every evening sorting unworked over or undistributed material, some of it more than thirty years old as you will see from the specimens I send you next time. This work is quite like a game and I enjoy it very much at the end of the day. There are several hundred bundles, most of which have never been touched for many years, and in the process of moving from the old Museum to the new one seven years ago, and in moving from the Museum to where we are now, these bundles have got nicely mixed and I am taking them just as they come.[33]

Although Jim had been doing this kind of work since joining the Survey in 1882, he now believed that he had more than served his apprenticeship. It was time for bigger things, particularly since he was now on his own without his father watching over his shoulder. He consequently told McConnell, in anticipation of the move back into the museum, that the botanical branch would require an assistant curator and herbarium assistant. 'I think it is due to me and the future of Canadian Botany that these appointments should be made as soon as possible,' he wrote, '... the best of the results of my many years of field and herbarium work should be prepared for publication with as little delay as may be. I hope that I will not be separated from the herbarium during my active working life and I honestly value the title "Curator of the National Herbarium," more than that of "Chief of the Biological Division." '34 The situation at the museum, however, remained discouraging; as McConnell advised Taverner and Anderson during a meeting in his office, 'it would be a long time before much could be done along museum lines, that everything would have to stand on its *economic* merits, until the country was in a way to get its war debts paid off.'35

Throughout this period, Jim spent a few weeks each year with his father and kept him informed of what was taking place in Ottawa. Whenever possible, he also tried to take his father to the field for short periods; in fact, Jim kept his camping gear at Sidney ready to be shipped whenever he planned to collect.36 In 1915, the trio of the two Macouns and Spreadborough, in the field together for the first time since 1890, collected halfway up the east coast of Vancouver Island to Comox. The following year, Jim, with Spreadborough and Young as his two assistants, worked along the Pacific Great Eastern rail line from Lilloet to Squamish, where they were later joined by the Professor for a few weeks at Howe Sound. The next three summers were devoted to an intensive botanical survey of the Jasper Park region. At the conclusion of each field season, however, Jim returned to Vancouver Island with Spreadborough for a quick visit with his father before heading back east.

It was while collecting along the Grand Trunk Pacific line from Jasper to Prince Rupert in 1919 that Jim fell seriously ill. Upon his return to Ottawa he tried to catch up on the summer's accumulation of office work, but his condition steadily deteriorated. Exploratory surgery in early November revealed that he was suffering from an advanced cancer. Following the operation he rallied, thanks to his strong constitution. Family and friends were warned, however, that it was only a matter of weeks.37 He died on 8 January 1920 at the age of fifty-seven. James Macoun's death, although expected, came as a great shock. The genial, unassuming naturalist had a great number of friends within government ranks and Ottawa circles, and his funeral was attended by a wide cross-section of mourners – including the man who had been Liberal minister of agriculture at the time of his controversial Peace River report. His co-workers at the museum were particularly devastated. 'We miss him greatly,' Taverner confided to a friend: 'He was a bigger man than generally realized. He had potentialities for almost anything, with a little more energy or ambition he would have been one of the big names in Canada. Personally I feel that I have lost one of my best friends, one upon whose judgment and willingness to assist I could rely under any circumstances. Departmentally he is irreplaceable. Most of what he knew died with him.'[38] Anderson, who had come to know Jim for only a brief period, was equally distressed. In breaking the news to William Spreadborough, he probably captured the feelings of many when he wrote, 'You know how it is – "Jimmie" Macoun was my best friend in Ottawa.'[39]

Following Jim's death, Taverner asked Spreadborough to help him survey the bird life of the southern prairies. Jim, however, had been like a brother to the field assistant – always looking out for his interests – and when he died, Spreadborough's heart for field work died with him. He therefore politely but steadfastly refused Taverner's repeated invitations to accompany him to the field and would only provide information on the best places to collect. Spreadborough promised to keep in touch but was last heard from in February 1922 when he was performing various outside jobs for the Esquimalt municipality. It was consequently not until two years after the event that Taverner learned of Spreadborough's tragic 1931 death on the day that he retired. 'The feeling of uselessness was too much for him,' Taverner related to Fleming, 'and he hanged himself in his little workshop leaving a note for his wife that they had not saved enough for two, rather than live on in poverty, it was better that he should pass out. Too bad, poor old Spreadborough.'[40]

The Professor, on learning of his son's death, lapsed into a severe depression and it was not until early March that he began to function again. 'Jim's death was an awful wrench,' he wrote to Dr M.O. Malte, a Department of Agriculture botanist who had worked quite extensively with his son in the Ottawa region. 'We had worked together for 40 years and all I had been doing was to build for him in the future. All this blasted in a day nearly finished me but I am in excellent health now but my botanical interest is dead ... My time here may be long or short but my work is done.'[41] Shortly thereafter, Dr Anderson, the new biological division chief, approached Macoun to inquire what should be done about the botany branch now that the government had finally vacated the museum. The Professor had earlier suggested to McConnell during his son's illness that Dr Malte should be appointed herbarium curator and he now repeated this wish. He was quite adamant on this point. 'My feelings regarding the proper care of the herbarium is [sic] like a mother to her only child,' he flatly told Anderson, 'and the appointment of a nonentity to take charge would almost kill me after having spent the best years of our life in building it up.'[42] Apart from this recommendation, Macoun had no desire to interfere in Ottawa matters in any other way. He turned over his Sidney herbarium to the provincial museum and resigned himself to his new honorary title as 'Naturalist on Vancouver Island.'[43] He also gave up on his autobiography, which he had recently commenced at Jim's urging, and it fell to his other son, William, to complete the book after his death.[44]

On 5 July 1920, in an unsteady hand, the Professor wrote his last letter to Dr Anderson at the biological division. Knowing that Dr Malte was to be in charge of the herbarium, he said that he could now 'die happy knowing all was well with the museum.' He then spoke of the number of recent deaths in Survey ranks and mused, 'all we can do is to look around our circle and ask who is next.'[45] As his heart steadily weakened, Macoun became more restricted in his movements and was eventually confined to bed. In the early morning hours of Sunday, 18 July 1920, he died peacefully in his sleep in his ninetieth year and was buried in Patricia Bay Cemetery. After the death of his wife in February of the following year, their bodies were taken back to Ottawa by train to be interred next to that of their son Jim in the Macoun plot in Beechwood cemetery. Macoun's death was marked by a number of lavish tributes in scientific journals and Canadian newspapers. The Ottawa Citizen, for example, carried a front-page obituary notice and then, two weeks later, a lengthy editorial that eulogized the late professor as 'a true citizen of Canada, a distinguished public servant with a vision of the country's greatness, and a capable and zealous investigator in the cause of science.'[46] What would have been equally gratifying for Macoun, however, was the knowledge that as of 1 December 1920, the Victoria Memorial Museum, including the biological division, was made a separate and distinct branch of the Department of Mines. Seven years later, on 5 January 1927, the Mackenzie King government finally declared the museum to be the National Museum of Canada – something the Macouns and Spreadborough had devoted the better part of their lives to realizing. The years of frustration were not over, though, for the Geological Survey retained its responsibility for the museum's budget and tended to keep the institution in a subordinate position. Taverner, Anderson, Patch, and Malte would chafe under many of the same problems that had earlier plagued Macoun and his son.[47] When, however, the National Museum was finally given the chance to live up to the expectations of its title following the Second World War, it owed much of its future success to a young Irish immigrant with a passion for the natural world.

Conclusion

On 20 December 1924, Dr M.O. Malte wrote to the deputy minister of mines, Charles Camsell, about securing possible assistance in the Dominion herbarium. The new chief botanist's request was nothing new – John Macoun and later his son, James, had regularly sought such help since the turn of the century. In support of his request, however, Malte included a candid assessment of the state of the herbarium and what lay ahead of him. He reported that approximately one-quarter of the herbarium specimens had to be renamed because of changes in botanical nomenclature adopted twenty-nine years earlier in Vienna. He also noted that the holdings were 'far from complete ... much more collecting will have to be done in order to make the herbarium truly national in character and extent.'[1] To support this contention, Malte explained to the deputy minister that recent botanical surveys had turned up species that were either new to the flora of Canada or not represented in the Dominion herbarium. He also cited the urgent need to prepare a flora of Canada, noting that the British Association for the Advancement of Science, meeting in Toronto in 1923, was shocked to learn that students of Canadian botany were still largely dependent on American literature.

These problems, as outlined by Malte, were a direct legacy of John Macoun's long tenure at the Geological Survey of Canada. In the latter part of the nineteenth century, natural history had begun to give way to the new science of biology. Investigators of the natural world were no longer expected to be all-round generalists but rather specialists in one particular field, be it birds, reptiles, or fungi. Taking up the Darwinian challenge, they also moved beyond the traditional naturalist's emphasis on accumulation and description in favour of studying the anatomy and physiology of organisms. In short, the problem-solving, economic-oriented approach that had characterized field work in the past was being replaced by pure, comprehensive research.

Macoun's work at the Survey went against this trend. When the Professor assumed his duties as Dominion botanist in 1882, he was one of the most renowned scientific figures in Canada. Over the preceding decade, he had used his formidable skills as a field naturalist to portray western Canada as an agricultural Eden. The popular success of this work, in particular the favour it found among federal politicians, had a lasting influence on Macoun's subsequent career at the Survey. From the time he took up his duties, his work was never questioned. It was simply assumed that he was engaged in the same kind of field work that he had performed in western Canada in the 1870s. Besides, whenever the assignment arose to determine a region's potential on the basis of its natural life, he saw it as a fundamental part of his natural history work and always returned brimming with optimism. He knew full well that Ottawa expected positive, practical information of immediate economic value; his son's sorry experience over his 1904 Peace River report was a case in point. Macoun's sweeping resource assessments, however, were not simply done to please his political masters. Unlike his contemporaries at the Survey during the Macdonald-Laurier years, his enthusiasm knew no bounds. He had a great faith in Canada's destiny and an even greater faith in his role in revealing that destiny. He was convinced that the young Dominion with its great resource heritage would soon become the home of a superior civilization. 'We have more than half a continent, and if we can raise first-class wheat,' he told the Commons Committee on Agriculture and Colonization in one of his frequent outbursts of hyperbole, 'certainly we ought to raise first-class men.'[2] Such blind optimism created the false impression that Canada's frontiers could be settled and developed with relative ease. In more general terms, it also served to substantiate the widely held belief that humankind was a beneficiary rather than an integral part of nature and that Canada's resource wealth was not only inexhaustible but made 'useful' through exploitation.

Macoun's view of the purpose of nature study also affected the kind of natural history that was pursued under Survey auspices. As a committed field naturalist who had developed an intimate knowledge of Canada's varied biological life and its habitat through firsthand contact, the Professor might have been expected to have shown some interest in Darwin's theory of natural selection. He had no time, however, for Darwin's thesis nor, for that matter, for the growing specialization within the discipline. He believed that a naturalist should be a kind of jack-of-all-fields whose role was to assemble as complete and accurate an inventory of God's wondrous bounty as possible. This obsession with enumerating Canada's natural life, with particular emphasis on the discovery of species new to science, meant that the Survey's natural history endeavours remained rooted in the early nineteenth century; it was part of a bygone era. Whereas the work of other natural scientists at similar institutions in Europe and the United States was increasingly focused on a particular field or genus, Macoun's duties, at his request, went in the opposite direction – from Dominion botanist to Survey naturalist. This 'backward' step, however, suited Canadian parliamentarians, concerned that the Survey was concentrating too much on pure science and was getting away from its original function of aiding in the development of the mineral resources of the country. They understood and appreciated the kind of field work that Macoun performed. Canada's natural life was simply another resource.

Macoun's work-load was to have its costs and benefits during his thirty-year Survey career; it was a double-edged sword. His enthusiasm, tenacity of purpose, and, above all, egotism created the false impression that he, his son, and Spreadborough, with occasional assistance from other Survey officers, could assemble a representative collection of Canada's biological life. The various Survey directors of the period never gave any serious thought to expanding the natural history staff, even after Macoun had been promoted in 1887 and given responsibility for collecting and cataloguing Canada's animal as well as plant life. In reality, however, the annual field outings were little more than haphazard sweeps through a region to gather whatever was chanced upon. It was a hit-and-miss exercise, as evidenced by Malte's comments about the gaps in the herbarium holdings. With the emphasis on making large collections in all areas, the specimens themselves were generally poorly documented (and in the case of vertebrates, poorly preserved), and field collections sometimes sat around for several years before they were gone over; there are still unopened bundles from the Macoun era in the basement of the botany division of the National Museum of Natural Sciences.

By concentrating most of his energies on field work, Macoun also had limited time to devote to detailed study and publication. The Professor believed it was more practicable, given his field of inquiry, to concentrate on what he did best – collecting – and leave it to others, in particular American specialists, to work up his specimens. 'While we may lose some kudos by following this plan,' Jim once explained in defence of their failure to produce a flora of Canada, 'we certainly do a great deal to add to our knowledge of Canadian botany.'[3] As collectors pure and simple, however, they were out of touch with the kind of biological research that was occurring at universities and other similar institutions. The biological division was essentially a misnomer, and it was not until the mid-twentieth century that the new science of biology found a home in the Victoria Memorial Museum. The heavy dependence on outside assistance also meant that Macoun, his son, and Spreadborough were little more than 'hewers of wood and drawers of water.' They supplied the field data and specimens for American research and publication on the biological life of the North American continent.

On a more positive note, even though John Macoun's original appointment was a political reward for his work in western Canada, he treated the position as a dream fulfilled. In fact, without the Professor's profound belief in himself and the importance of his work, it is unlikely that the Geological Survey would have been engaged in natural history to the extent that it was. During his tenure, not only did the geological duties of the Survey take precedence, but the agency itself was regularly criticized for not doing enough to promote the mining interests of the country. Macoun, however, did whatever was necessary to see that his studies did not suffer. Tackling his work with missionary zeal, he was willing to adopt any tactic, including writing letters to newspapers, calling on political friends, or scheming with fellow scientists, to see that he was sent to the field every season, even in those years when the Survey was financially strapped. And in the long run he upheld the natural history duties of the Survey in such a fashion that they came to be recognized as a legitimate concern.

Macoun's wide-ranging field activity also figured in the establishment of a national museum. At the end of each field season, he would return to Ottawa heavily laden with specimens. This material, together with that gathered by his son Jim, Spreadborough, and other officers, eventually reached such proportions that the Survey found itself in possession of what was generally considered a 'national collection,' but with no room to store these objects safely, let alone display them. In the campaign to secure a new Survey building that would be a fitting repository for these so-called national treasures, Macoun's field efforts were regularly used to justify such a structure. When the Victoria Memorial Museum was finally occupied, the material that had been assembled since the early 1880s, despite its shortcomings, laid the foundation upon which future museum biologists would build.

Macoun's greatest legacy was his ability as a field naturalist – something that Malte alluded to in his letter to Camsell: 'May I be permitted to say, quite frankly,' he concluded, 'that it is a physical impossibility for one man single-handed to cover a territory as large and as varied ... This was also presumably realized by Department when I was appointed Chief Botanist, and not Botanist pure and simple.'[4] This was precisely what the Professor attempted to do during his Survey career. He collected as widely and thoroughly as possible in the field, usually working from dawn to dusk. Few hardships or obstacles deterred him. In the process, he developed such an unrivalled knowledge of Canada's natural life that he could recognize new forms at sight and discovered several species or subspecies new to science, many of which were named after him.[5] What was most amazing about his collecting prowess, however, was the range of territory that he covered; he logged thousands of miles. His collections, moreover, were not only the first extensive ones made in particular areas, but in many instances, were assembled before the natural environment was disturbed by man. It is not too exaggerated to say that John Macoun tried almost single-handed to roll back the natural history frontiers of Canada. This reputation serves today as an inspiration to those who follow in his footsteps, professionals and amateurs alike. As the Professor's good friend A.H. Mackay said to the 1922 meeting of the Royal Society of Canada in memorializing the recently deceased dean of Canadian naturalists: 'No great explorer of Nature exhausts the field. He only opens to view many other boundless areas.'[6]

Picture Credits

National Archives of Canada: John and Ellen Macoun c55358; John
  Macoun 1902 PA33784; James Macoun on Low expedition 1885
  PA37925; Selwyn PA25677; Dawson PA25521; Victoria Memorial
  Museum 1907 PA42281
National Museums of Canada: Macoun in Maritimes 1890s J2809;
  James Macoun to Hudson Bay 1910 J5542; Macoun in Victoria
  Museum 1911 31874; Macoun and Taverner J5535; Taverner
  44459; Anderson in Mackenzie Delta 38748; Geological Survey's
  headquarters 109124; Victoria Museum 1912 18806A; Victoria
  Museum 1927 69308
W. Spreadborough, Bracebridge, Ontario: Spreadborough
Whyte Museum of the Canadian Rockies: Kicking Horse Pass 1884
  NA66-2292; Hiking party 1912 PA390-33
Geological Survey of Canada: Brock 201772; Bell 68775; Macoun,
  Young, and Spreadborough 1909 202423; Exhibition Hall, Survey
  museum 201736-A
Eleanor Sanderson, Victoria, British Columbia: W.T. Macoun with
  parents at Sidney; gathering at the A.O. Wheeler home; Ellen
  Macoun and Miss W.H. Fatt

Note on Sources

Most studies of John Macoun focus on his field work in western Canada in the 1870s and his reversal of Palliser's and Hind's conclusions about the region's potential. In fact, Macoun himself devoted almost two-thirds of his *Autobiography* (Ottawa 1922) to this topic. His thirty-year career as Dominion botanist and Survey naturalist has consequently been largely overlooked.

The Manuscript Division of the National Archives of Canada has a small collection of John Macoun papers, consisting of his expedition notebooks for the 1870s. References to Macoun and his work can also be found in the papers of some of his scientific contemporaries (Sandford Fleming; Robert Bell; R.M. Anderson) and political connections (Sir John A. Macdonald; Sir Wilfrid Laurier; Sir Clifford Sifton), and in various record groups (Interior; Geological Survey; Public Works; Canadian Government Exhibition Commission; Privy Council) of the Government Archives Division, NAC. There are only a few private Macoun papers, mostly letters to his wife during the 1870s, in the possession of an American descendant.

The National Museums of Canada Library in Ottawa has nine volumes of Macoun Letterbooks (for John and his son James) for the period from November 1884 to April 1902. Additional correspondence, mostly botanical in nature, was held by the Botany Division of the National Museum of Natural Sciences and has recently been transferred to the National Archives. All Macoun's botanical field notebooks, as well as those of his son, remain in the possession of the Botany Division. Correspondence dealing with birds and mammals was appropriated by Macoun's successors, Percy Taverner and R.M. Anderson respectively, and is located today at the Vertebrate Zoology Division of the National Museum of Natural Sciences. The Taverner correspondence at the Royal Ontario Museum contains references to the Macouns and to William Spreadborough, and Macoun material can also be found in the various depositories that hold the papers of his correspondents. The Gray Herbarium of Harvard University and the Royal Botanic Gardens, Kew, in London, for example, have a number of Macoun letters for the 1870s.

The activities of the natural history branch (later biological division) in the field and office are chronicled in the annual Geological Survey reports. These should be read in conjunction with the Survey director's summary reports. Another valuable source is the transactions of the Ottawa Field-Naturalists' Club; many officers of the Geological Survey and other government officials were active members of the club.

Macoun's ideas on western Canada's potential and Canada's future as a great agricultural nation were brought together in his extremely popular *Manitoba and the Great North-West* (Guelph 1882). He also published a three-volume *Catalogue of Canadian Plants* (Ottawa 1886; 1890; 1901) and was co-author with his son of the *Catalogue of Canadian Birds* (Ottawa 1900; 1904).

A number of secondary sources should be consulted for context, as well as for perspective on Macoun's work. Carl Berger, *Science, God and Nature in Victorian Canada* (Toronto 1983), and Brian McKillop, *A Disciplined Intelligence: Critical Inquiry and Canadian Thought in the Victorian Era* (Montreal 1979), discuss Canadian natural science and the Darwinian challenge. A critical assessment of Macoun's favourable evaluation of western Canada's agricultural capabilities is provided by Doug Owram, *Promise of Eden: The Canadian Expansionist Movement and the Idea of the West, 1856–1900* (Toronto 1980). The activities, concerns, and problems of the Geological Survey during Macoun's tenure are described in Morris Zaslow, *Reading the Rocks: The Story of the Geological Survey of Canada, 1842–1972* (Toronto 1975). Finally, the state of natural history research in Canada before and after Macoun is explored in Suzanne Zeller, *Inventing Canada: Early Victorian Science and the Idea of a Transcontinental Nation* (Toronto 1987), and J. Foster, *Working for Wildlife: The Beginning of Preservation in Canada* (Toronto 1978).

For more detailed bibliographical information, the reader should consult the footnotes.

Notes

ABBREVIATIONS USED IN NOTES

GHHU   Gray Herbarium of Harvard University
GSC   Geological Survey of Canada
MRB   McGill Rare Book Library
NAC   National Archives of Canada
NMC   National Museums of Canada Library
NMNS   National Museum of Natural Sciences
PABC   Provincial Archives of British Columbia
QUA   Queen's University Archives
RKG   Royal Botanic Gardens, Kew
ROM   Royal Ontario Museum
SIA   Smithsonian Institution Archives
USASC   University of Saskatchewan Archives and Special Collections
UT   University of Toronto Library

INTRODUCTION

1 A.B. McKillop, *A Disciplined Intelligence: Critical Inquiry and Canadian Thought in the Victorian Era* (Montreal 1979), 93–134
2 D.R. Owram, *Promise of Eden: The Canadian Expansionist Movement and the Idea of the West* (Toronto 1980), 59–78
3 For a discussion of this issue, see V. de Vecchi, 'Science and Government in the Nineteenth Century' (PHD diss., University of Toronto 1978)
4 'You are right in pronouncing my name *Macown*. Scotchmen call me Macoon which is wrong.' NMNS, Botany Division, J. Macoun to C. Dewey, 12 Dec. 1865. The material on Macoun's life up until 1872 is

largely based on *Autobiography of John Macoun: Canadian Explorer and Naturalist* (Ottawa 1922).

5 NMNS, Botany Division, J. Macoun to C. Dewey, 9 November 1865

6 J. Macoun, 'Dr. Fletcher as a Naturalist,' *Ottawa Naturalist*, 22, no. 10 (January 1908–9), 214.

7 See A. Prentice, *The School Promoters: Education and Social Class in Mid-Nineteenth Century Upper Canada* (Toronto 1977); J. Love, 'The Professionalization of Teachers in Mid-Nineteenth Century Upper Canada,' in N. McDonald and A. Chaiton, eds, *Egerton Ryerson and His Times* (Toronto 1978), 109–28

8 NMCL, Macoun Letterbooks, vol. 7, 78, J. Macoun to A. Brody, 4 January 1894

9 Macoun, *Autobiography*, 40

10 NMCL, Macoun Letterbooks, vol. 6, 63, J. Macoun to J.R. Bagnall, 24 September 1892

11 G.E. Boyce, ed., *Hutton of Hastings: The Life and Letters of William Hutton, 1801–1861* (Belleville 1972), 236–45

12 NMCL, Macoun Letterbooks, vol. 8, 72, J. Macoun to J.J. Morgan, 21 March 1895. For a more detailed description of the object-lesson method and its place in Ontario education history, see G. Killan, *David Boyle: From Artisan to Archaeologist* (Toronto 1983), 20–39. Macoun's career and activities bear a striking resemblance to those of Boyle, Canada's first professional archaeologist.

13 RKG, North American and South American Letters, vol. 65, J. Macoun to W.J. Hooker, 14 October 1864

14 NMNS, Botany Division, J. Macoun to P. Cox, nd (January 1907?)

15 For a good general discussion of this issue, see J.C. Greene, *Death of Adam: Evolution and Its Impact on Western Thought* (Ames 1959).

16 M. Ruse, *The Darwinian Revolution: Science Red in Tooth and Claw* (Chicago 1979), 160–201

17 W. Coleman, *Biology in the Nineteenth Century: Problems of Form, Function and Transformation* (Cambridge 1977), 1–15

18 A.H. Dupree, *Asa Gray* (New York 1968), 386–7

19 McKillop, *A Disciplined Intelligence*, 99–100; C. Berger, *Science, God and Nature in Victorian Canada* (Toronto 1983), 70. Darwin's ideas were a topic of lively debate by some of Ontario's local natural history societies. See, for example, Killan, *David Boyle*, 59–61.

20 Berger, *Science, God, and Nature in Victorian Canada*, 13–16

21 *Ibid.*, 6; 9; R.A. Jarrell, 'Science as Culture in Victorian Toronto,' *Atkinson Review of Canadian Studies*, 1, no. 1 (Fall 1983), 5

22 McKillop, *A Disciplined Intelligence*, 23, 59–61, 95–8

23 J. Barnston, 'General Remarks on the Study of Nature, With Special reference to Botany,' *Canadian Naturalist and Geologist*, 2 (1857), 34

24 Macoun, *Autobiography*, 99

25 *Ibid.*, 38–9; L.E. Wilson, ed., *Sir Charles Lyell's Scientific Journals on the Species Question* (New Haven 1970), xxxi; C.C. Gillespie, *Genesis and Geology* (New York 1959), ch. 4

26 NMNS, Botany Division, J. Macoun to C. Dewey, 6 November 1865

27 RKG, British North America Letters, vol. 195, J. Macoun to J.D. Hooker, 7 August 1866

28 S.F. Cannon, *Science in Culture: The Early Victorian Period* (New York 1978), 73–110

29 Macoun, *Autobiography*, 40

30 J. Macoun, 'The Capabilities of the Prairie Lands of the Great North-West, as Shown by Their Fauna and Flora,' *Ottawa Field-Naturalists' Club Transactions*, 3, no. 2 (1881), 38

31 S. Zeller, *Inventing Canada: Early Victorian Science and the Idea of a Transcontinental Union* (Toronto 1987), 230–4

32 RKG, North American and South American Letters, vol. 65, J. Macoun to W.J. Hooker, 4 November 1863

33 *Ibid.*, 14 October 1864

34 NMNS, Botany Division, J. Macoun to C. Dewey, 7 August 1865

35 *Ibid.*, 6 December 1865

36 QUA, A.T. Drummond Papers, J. Macoun to A.T. Drummond, 6 October 1866

37 *Ibid.*, 19 March 1867

38 W.E.L. Smith, *Albert College, 1857–1957* (Kingston 1957), 14–20; R. Cook, *The Regenerators: Social Criticism in Late Victorian English Canada* (Toronto 1985), 20–4, 193

39 J.S. Pringle, 'John Macoun's Academic Degrees,' *The Plant Press*, 5, no. 1, 19–20

40 GHHU, Macoun Correspondence, J. Macoun to A. Gray, 21 November 1874

41 RKG, British North America Letters, vol. 195, J. Macoun to J.D. Hooker, 8 October 1867

CHAPTER 1 Getting Aboard

1 J. Macoun, *Manitoba and the Great North-West* (Guelph 1882), 219

2 J. Macleod Papers, J. Macoun to R. Kingman, 9 June 1918

3 For a discussion of pre-1870 western agriculture, see A.S. Morton, *A History of Prairie Settlement* (Toronto 1938); F.G. Roe, 'Early Agricul-

ture in Western Canada in Relation to Climatic Stability,' *Agricultural History*, 26, 3 (July 1952), 104–23

4 E.C. Hope, 'Weather and Crop History in Western Canada,' *Canadian Society of Technical Agriculturalists Review*, 16 (March 1938), 349–50

5 J. Warkentin, 'The Desert Goes North,' in B.W. Blouet and M.P. Lawson, eds, *Images of the Plains: The Role of Human Nature in Settlement* (Lincoln 1975), 151–2, 154–7; J. Warkentin, 'Steppe, Desert and Empire,' in A.W. Rasporich and H.C. Klassen, eds, *Prairie Perspectives* 2 (Toronto 1973), 113–21

6 Owram, *Promise of Eden: The Canadian Expansionist Movement and the Idea of the West* (Toronto 1980), 59–78

7 E.H. Oliver, ed., *The Canadian North-West*, vol. 2 (Ottawa 1915), 958 (Deed of Surrender)

8 G.M. Grant, *Ocean to Ocean: Sandford Fleming's Expedition through Canada in 1872* (Toronto 1873), 35

9 *Ibid.*, 23

10 *Ibid.*, 60

11 Macleod Papers, J. Macoun to E. Macoun, 1 August 1872

12 Bishop Alexandre Taché had outlined his views two years earlier in *Sketch of the North-West of America*, trans. (Montreal 1870).

13 M. McLeod, ed., *Peace River: A Canoe Voyage from Hudson Bay to Pacific by the Late George Simpson* (Ottawa 1872)

14 Macleod Papers, J. Macoun to E. Macoun, 28 August 1872

15 NAC, Manuscript Division, MG 24K, John Macoun Papers, 1872 field notebook, 14 October 1872

16 C. Tracie, 'Land of Plenty or Poor Man's Land: Environmental Perception and Appraisal Respecting Agricultural Settlement in the Peace River Country, Canada,' in Blouet and Lawson, eds, *Images of the Plains*, 120–1

17 Macleod Papers, J. Macoun to R. Kingman, 9 June 1918

18 See J.M. Thorington, 'The Cariboo Journal of John Macoun,' *Geographical Society of Philadelphia Bulletin*, 28, no. 3 (July 1930), 199–209

19 Macleod Papers, J. Macoun to E. Macoun, 1 August 1872

20 J. Macoun, 'Botanical Report, Lake Superior to Pacific Ocean,' in S. Fleming, ed., *Canadian Pacific Railway Report of Progress on the Explorations and Surveys up to January 1874* (Ottawa 1874), App. C, 65

21 *Ibid.*, 66

22 *Ibid.*, 67

23 NAC, Government Archives Division, RG 45, GSC, Directors' Letterbooks, vol. 74, 198, A.R.C. Selwyn to A.E. Meredith, 11 June 1872

24 *Canadian Journal*, 3, no. 17 (September 1858), 461–2

25 GSC, Directors' Letterbooks, vol. 74, 198, A.R.C. Selwyn to A.E. Meredith, 11 June 1872. In 1858, English botanist W.S.M. D'Urban assisted Logan during his survey of the Grenville, Quebec, region.

26 *Ibid.*, vol. 84, 139, A.R.C. Selwyn to R.H. Dana, 26 December 1889

27 M. Zaslow, *Reading the Rocks: The Story of the Geological Survey of Canada, 1842–1972* (Toronto 1975), 34, 101

28 GSC, Directors' Letterbooks, vol. 74, 198, A.R.C. Selwyn to A.G. Meredith, 11 June 1872

29 *Ibid.*, vol. 75, 158, A.R.C. Selwyn to J. Macoun, 27 April 1874

30 Macleod Papers, J. Macoun to E. Macoun, 28 August 1872

31 NAC, Manuscript Division, MG 29, Sandford Fleming Papers, J. Macoun to S. Fleming, 15 May 1874

32 GSC, Directors' Letterbooks, vol. 75, 205, A.R.C. Selwyn to J. Macoun, 10 June 1874

33 GHHU, John Macoun Correspondence, J. Macoun to Asa Gray, 21 November 1874, 9 January 1875

34 A. Birrell, 'Fortunes of a Misfit: Charles Horetzky,' *Alberta Historical Review*, 19, no. 1 (Winter 1971), 12–14

35 Macleod Papers, J. Macoun to E. Macoun, 23 April 1875

36 RKG, British North America Letters, vol. 195, J. Macoun to J.D. Hooker, 5 September 1876

37 Macoun (in his *Autobiography of John Macoun: Canadian Explorer and Naturalist* [Ottawa 1922], 111–16) and King (see M. Weekes, *Trader King* [Regina 1949], 143–8) offer differing accounts of this misadventure.

38 NAC, Manuscript Division, MG 29, Roderick MacFarlane Papers, vol. 1, 499–502, J. Macoun to R. MacFarlane, 19 December 1875

39 J. Macoun, 'Report on the Botanical Features of the Country Traversed from Vancouver to Carlton on the Saskatchewan,' *Geological Survey of Canada Report of Progress for 1875–1876* (Montreal 1876), 152, 155

40 Macoun Papers, Geological Notes on Peace River, 1875

41 *Ibid.*

42 MacFarlane Papers, vol. 1, 499–502, J. Macoun to R. MacFarlane, 19 December 1875

43 Macoun Papers, 1875 field notebook, 7 September 1875

44 *Ibid.*, 19 October 1875

45 GSC, Directors' Letterbooks, vol. 75, 719, A.R.C. Selwyn to J. Macoun, 18 December 1875; 748, A.R.C. Selwyn to J. Macoun, 29 December 1875

46 MacFarlane Papers, vol. 1, 499–502, J. Macoun to R. MacFarlane, 19 December 1875

47 Macoun, *Autobiography*, 132

48 See T. Rawlings, *The Confederation of the British North American Provinces* (London 1865), App. B, 212–16

49 G.S. Dunbar, 'Isotherms and Politics: Perceptions of the North-West in the 1850s,' in Rasporich and Klassen, eds, *Prairie Perspectives 2*, 80–101

50 Taché, *Sketch of the North-West of America*, 17

51 Macoun, 'Report on the Botanical Features ... from Vancouver to Carlton,' 154

52 Macoun, *Autobiography*, 134

53 PABC, Malcolm McLeod Papers, J. Macoun to M. McLeod, 11 March 1876

54 Canada, House of Commons, *Journals*, 1876, vol. 10, App. 8, 'Report of the Standing Committee on Agriculture and Colonization,' 28

55 *Ibid.*, 21

56 G.M. Dawson, *Report on the Geology and Resources of the Region in the Vicinity of the 49th Parallel, from Lake of the Woods to the Rocky Mountains* (Montreal 1875), 299

57 *Ibid.*, 319

58 *Ibid.*, 292, 299

59 *Ibid.*, 301

60 Warkentin, 'Steppe, Desert and Empire,' 123–4; J. Warkentin, ed., *The Western Interior of Canada* (Toronto 1969), 235

61 MUA, George Mercer Dawson Papers, J. Macoun to G.M. Dawson, 10 December 1875

62 Dawson, *Geology and Resources*, 301

63 J. Macoun, 'Sketch of that Portion of Canada between Lake Superior and the Rocky Mountains, with Special Reference to Its Agricultural Capabilities,' in S. Fleming, ed., *Report on Surveys and Preliminary Operations on the Canadian Pacific Railway up to January 1877* (Ottawa 1877), App. x, 335

64 *Ibid.*, 336

65 *Ibid.*, 334

66 Macoun, *Autobiography*, 158

67 Zaslow, *Reading the Rocks*, 123–4

68 Canada, House of Commons, *Debates*, 27 February 1877, 314

69 *Ibid.*, 316

70 See, for example, GSC, Directors' Letterbooks, vol. 78, 78 , A.R.C. Selwyn to J. Macoun, 11 December 1880.

71 A. Wilson, 'Fleming and Tupper: The Fall of the Siamese Twins, 1880,' in J.S. Moir, ed., *Character and Circumstance: Essays in Honour of Donald Grant Creighton* (Toronto 1970), 115–16

72 Macoun, *Autobiography*, 135

73 Canada, House of Commons, *Debates*, 10 May 1879, 1893
74 T. Spence, *The Saskatchewan Country of the North-West of the Dominion of Canada* (Montreal 1877), 26
75 J. Trow, *Manitoba and NorthWest Territories* (Ottawa 1878), 29
76 Fleming quoted in C.R. Tuttle, *Our North Land* (Toronto 1885), 312
77 Owram, *Promise of Eden*, 152
78 Macoun, *Manitoba and the Great North-West*, 264
79 F.G. Roe, 'Early Opinions on the "Fertile Belt" of Western Canada,' *Canadian Historical Review*, 27, no. 2 (June 1946), 149
80 Canada, House of Commons, *Sessional Papers*, 1880, vol. 3, no. 4, part ii, 'Report of the Department of the Interior for 1879,' 6
81 Fleming Papers, J. Macoun to S. Fleming, 20 May 1878
82 Macleod Papers, J. Macoun to E. Macoun, 10 June 1879
83 *Ibid.*, 12 June 1879
84 Hope, 'Weather and Crop History in Western Canada,' 351, 358
85 Macleod Papers, J. Macoun to E. Macoun, 19 June 1879
86 *Manitoba Free Press*, 21 November 1879, 1
87 J. Macoun, 'General Remarks on the Land, Wood and Water of the North-West Territories, from the 102nd to 115th Meridian and between the 51st and 53rd Parallels of Latitude,' in S. Fleming, ed., *Report and Documents in Reference to the Canadian Pacific Railway* (Ottawa 1880), App. 14, 240
88 *Ibid.*, 236–9
89 *Ibid.*, 239
90 *Ibid.*
91 J. Hector, 'On the Physical Features of the Central Part of British North America and on Its Capabilities for Settlement,' *Edinburgh New Philosophical Journal*, new series, 14, no. 2 (October 1861), 22
92 *Ibid.*, 10–11, 22
93 Dawson, *Geology and Resources*, 317
94 *Ibid.*, 283–4, 292–4
95 *Manitoba Free Press*, 20 November 1879, 1
96 Although Macoun never made any reference in his speech to the source of his ideas on the climate of the western interior, it would appear that he consulted J. Hurlbert, *The Climates, Productions and Resources of Canada* (Montreal 1872) and H.J. Coffin, *The Winds of the Globe* (Washington 1875). See Owram, *Promise of Eden*, 157–160.
97 J. Macoun, 'Notes on the Physical Phenomena of Manitoba and the North-West Territories,' *Canadian Journal*, 3rd series, vol. 1 (1879), 152
98 *Ibid.*, 155–9

99 C. Horetzky, *Some Startling Facts Relating to the Canadian Pacific Railway and the North-West Lands* (Ottawa 1880), 43
100 Birrell, 'Fortunes of a Misfit,' 22–4
101 Canada, House of Commons, *Debates*, 3 March 1880, 391
102 *Ibid.*, 15 April 1880, 1407–9; 1417
103 Macoun, *Manitoba and the Great North-West*, 612
104 Macleod Papers, J. Macoun to E. Macoun, 12 June 1880
105 *Ibid.*, 12 July 1880
106 *Ibid.*, 25 July 1880
107 Fleming Papers, J. Macoun to S. Fleming, 14 August 1880
108 *Ibid.*
109 J. Macoun, 'Extract from a Report of Exploration in the North-West Territories,' in Canada, House of Commons, *Sessional Papers*, 1881, vol. 3, no. 3, part i, 'Report of the Department of the Interior for 1880,' 22
110 *Ibid.*, 15
111 Macoun, 'General Remarks on the Land, Wood and Water,' 237
112 Macoun, 'Extract from a Report of Exploration,' 22
113 Macoun quoted in *Manitoba Free Press*, 7 April 1881, 1
114 H.N. Smith, 'Rain Follows the Plow: The Notion of Increased Rainfall for the Great Plains, 1844–1880,' *Huntington Library Quarterly*, 10, no. 1 (February 1947), 171–4
115 Macoun, *Manitoba and the Great North-West*, 294
116 Canada, House of Commons, *Debates*, 14 December 1880, 73
117 *Statutes of Canada*, 44 Victoria, 1880, ch. 1, 11
118 Macoun, 'General Remarks on the Land, Wood and Water,' 245
119 J. Macoun, 'Report of Exploration,' in Canada, *Sessional Papers*, 1882, vol. 8, no. 18, part i, 'Report of the Department of the Interior for 1881,' 80
120 Macoun, *Manitoba and the Great North-West*, 219
121 *Ibid.*, 610
122 *Ibid.*, 364
123 L.G. Thomas, 'Associations and Communications,' *Historical Papers 1973*, 8
124 For a discussion of this issue, see W.A. Waiser, 'A Willing Scapegoat: John Macoun and the Route of the CPR,' *Prairie Forum*, 10, no. 1 (Spring 1985), 65–81
125 Canada, House of Commons, *Debates*, 27 February 1877, 314

CHAPTER 2 Plants and Politics

1 R.C. Brown, 'The Doctrine of Usefulness: Natural Resource and National

Park Policy in Canada, 1887–1914,' in J.G. Nelson, ed., *Canadian Parks in Perspective* (Montreal 1970), 55

2 T. Levere, 'What Is Canadian about Science in Canadian History?' in R.A. Jarrell and N.R. Ball, eds, *Science, Technology and Canadian History* (Waterloo 1980), 20

3 V. de Vecchi, 'Science and Government in the Nineteenth Century' (PHD diss., University of Toronto 1978), 8–10, 164–5, 222–3

4 *Ibid.*, 168

5 J. Macoun, *Manitoba and the Great North-West* (Guelph 1882), 214

6 USASC, Shortt Collection, J. Macoun to Deputy Minister of the Interior, 12 May 1880

7 NAC, Government Archives Division, RG 15, Department of Interior, vol. 554, f. 165895, J. Macoun to M. Bowell, 7 March 1882

8 *Ibid.*, J. Macoun to L. Russell, 18 October 1882

9 *Ibid.*, notation on reverse side of draft of order-in-council, nd; notation by A.R.C. Selwyn on outside of file, 26 October 1882

10 NAC, Government Archives Division, RG 45, GSC, Directors' Letterbooks, vol. 78, 86, A.R.C. Selwyn to J. Macoun, 20 December 1880

11 D.P. Penhallow, 'A Review of Canadian Botany from the First Settlement of New France to the 19th Century,' *Transactions of the Royal Society of Canada*, series 1, vol. 5 (1887), sect. iv, 45–61; Frère Marie-Victorin, 'Canada's Contribution to the Science of Botany,' in H.M. Tory, ed., *A History of Science in Canada* (Toronto 1939), 35–40

12 George Lawson, 'Remarks on the Present State of Botany in Canada, and the Objects to Be Attained by the Establishment of a Botanical Society,' in G.F.G. Stanley, ed., *Pioneers of Canadian Science* (Toronto 1966), 131

13 W.K. Lamb and T.W.M. Cameron, 'Biologists and Biological Research Since 1864' in Stanley, ed., *Pioneers of Canadian Science*, 36

14 R. Duchesne, 'Science et société coloniale: les naturalistes du Canada français et leurs correspondants scientifiques (1860–1900),' *H.S.T.C. Bulletin*, 5, no. 2 (May 1981), 102–13; B. Boivin, 'Canadian Botany since Hooker (1840–1979),' unpublished article in author's possession, 1–2

15 J. Macoun, *Catalogue of the Phaenogamous and Cryptogamous Plants of the Dominion of Canada* (Belleville 1878)

16 A.H. Dupree, *Asa Gray* (New York 1968), 391

17 NMNS, Botany Division, A. Gray to J. Macoun, 13 October 1878

18 GHHU, John Macoun Correspondence, J. Macoun to A. Gray, 23 November 1881

19 NMNS, Botany Division, J. Macoun to E.G. Britton, 21 June 1891

20 GHHU, Macoun Correspondence, J. Macoun to S. Watson, 17 April 1884

21 *Ibid.*, J. Macoun to A. Gray, 1 April 1884

22 RKG, British North America Letters, J. Macoun to J.D. Hooker, 20 March 1883

23 Canada, House of Commons, *Journals* (1883), vol. 17, App. c, 'Report of the Select Standing Committee on Agriculture and Colonization,' 69

24 W. Eggleston, *The Queen's Choice: A Story of Canada's Capital* (Ottawa 1961), 140–3

25 de Vecchi, 'Science and Government in the Nineteenth Century,' 96

26 NMNS, Botany Division, E.E.T. Seton to J. Macoun, 26 May 1884

27 NMCL, Macoun Letterbooks, vol. 9, 124, J. Macoun to H.H. Gowan, 5 May 1896

28 T.C. Weston, *Reminiscences among the Rocks* (Toronto 1889), 177

29 See, for example, Duchesne, 'Science et société coloniale.'

30 GHHU, Macoun Correspondence, J. Macoun to A. Gray, 17 August 1882

31 NMCL, Macoun Letterbooks, vol. 4, 313, J. Macoun to G.N. Best, 8 January 1891

32 *Ibid.*, vol. 1, 5, J. Macoun to T. Morong, 8 February 1887

33 Ottawa *Citizen*, 10 April 1883, 4

34 *Ibid.*, 13 April 1883, 4

35 M. Zaslow, *Reading the Rocks: The Story of the Geological Survey of Canada, 1842–1972* (Toronto 1975), 136

36 J. Macoun, *Autobiography of John Macoun: Canadian Explorer and Naturalist* (Ottawa 1922), 207–9

37 NAC, Manuscript Division, MG 26A, John A. Macdonald Papers, vol. 260, 118675–7, L. Russell to J.A. Macdonald, 13 April 1883

38 House of Commons, *Debates*, 9 May 1883, 1088–9

39 Macdonald Papers, vol. 253, 114639, T. Macfarlane to J.A. Mousseau, 28 January 1879

40 House of Commons, *Debates*, 27 February 1877, 312

41 Macdonald Papers, vol. 253, 114656, T. Macfarlane to J.A. Mousseau, 16 February 1880

42 House of Commons, *Debates*, 26 April 1882, 1184

43 Zaslow, *Reading the Rocks*, 135; de Vecchi, 'Science and Government in the Nineteenth Century,' 164–9

44 Macdonald Papers, vol. 313, 141940–3, Geological Survey 1881

45 de Vecchi, 'Science and Government in the Nineteenth Century,' 222–3

46 Macdonald Papers, vol. 393, 188081–2, J. Macoun to J.A. Macdonald, 22 May 1883

47 Boivin, 'Canadian Botany since Hooker (1840–1879),' 4

48 GSC, vol. 16, f. 3, miscellaneous correspondence

49 NAC, Manuscript Division, MG 29, Robert Bell Papers, vol. 29, Subject

Files, Geological Survey, 'Bystander' to Editor of Toronto *Mail*, 11 April 1883

50 GSC, vol. 16, f. 3, miscellaneous correspondence
51 H.M. Ami, 'Sketch of the Life and Work of the Late Dr. A.R.C. Selwyn,' *The American Geologist*, 31, no. 1 (January 1903), 1–21
52 GSC, vol. 10, f. 2, miscellaneous correspondence, A.R.C. Selwyn to J.A. Macdonald, 1882
53 Zaslow, *Reading the Rocks*, 134–5
54 Macdonald Papers, vol. 249, 112588–9, D.L. Macpherson to J.A. Macdonald, 20 April 1883
55 Zaslow, *Reading the Rocks*, 56–7
56 T.C. Manning, *Government in Science: The U.S. Geological Survey 1867–1894* (Louisville 1967), 58–9, 74, 122–3
57 *Report of the Select Committee Appointed by the House of Commons to Obtain Information as to Geological Surveys* (Ottawa 1884), 49
58 *Ibid.*, 87
59 *Ibid.*, 84
60 *Ibid.*, 186
61 *Ibid.*, 185
62 *Ibid.*
63 House of Commons, *Debates*, 9 April 1884, 1451
64 *Ibid.*
65 Macdonald Papers, vol. 405, 195200–1, J.W. Dawson to J.A. Macdonald, 4 June 1884
66 GSC, vol. 79, 454, A.R.C. Selwyn to A.B. Perry, 5 April 1884
67 Department of Interior, vol. 320, f. 74880, R. Bell to Minister of the Interior, 29 April 1884
68 NAC, Manuscript Division, MG 30, Otto J. Klotz Papers, diary, 17 February 1886
69 *Report of the Select Committee ... Geological Surveys*, 8
70 *Ibid.*
71 *Ibid.*, 11
72 GSC, Directors' Letterbooks, vol. 79, 480, A.R.C. Selwyn to H. Merritt, 17 April 1884
73 Macdonald Papers, vol. 321, 144858, A.R.C. Selwyn to J.A. Macdonald, 2 March 1885
74 House of Commons, *Debates*, 12 July 1885, 3347–8
75 *Ibid.*, 3348
76 Zaslow, *Reading the Rocks*, 140–2
77 For a list of geographical features named after Macoun, see *Canoma*, 7, no. 1 (July 1981), 24–5

78 Macoun, *Autobiography*, 215–16
79 T.H. Levere, 'The British Association Goes West: Montreal 1884,' *Transactions of the Royal Society of Canada*, series 4, vol. 20 (1982), 489–97; V. de Vecchi, 'The Pilgrim's Progress, the BAAS, and Research in Canada: From Montreal to Toronto,' in *ibid.*, 526–8
80 J.P. Sheldon, *A Travers de Canada en compaignie de l'Association Brittannique* (Ottawa 1884), 25
81 NMNS, Botany Division, N.A. Martin to J. Macoun, 1 October 1884
82 Zaslow, *Reading the Rocks*, 133
83 NAC, Government Archives Division, RG 72, Canadian Government Exhibition Commission, vol. 37, f. 79, J.P. Sheldon to C. Tupper, 11 November 1885; f. 105, J.D. Hooker to C. Tupper, 25 November 1885
84 NMNS, Botany Division, W. Fream to J. Macoun, 3 December 1885
85 *Ibid.*, H.T. Mennell to J. Macoun, 2 December 1885
86 *Ibid.*, W.B. Cheadle to J. Macoun, 11 January 1886
87 Department of Interior, vol. 49, f. 105891, A.R.C. Selwyn to T. White, 1 February 1880
88 NMNS, Botany Division, G.B. Longstaff to J. Macoun, 11 October 1884
89 NMCL, Macoun Letterbooks, vol. 1, 387–8, J. Macoun to W.C. Van Horne, 29 March 1888
90 'Report of Sir Charles Tupper on the Canadian Section of the Colonial and Indian Exhibition, 1886,' in Canada, House of Commons, *Sessional Papers*, 1887, vol. 20, no. 12, 'Report of the Minister of Agriculture,' 70
91 NMCL, Macoun Letterbooks, vol. 1, 39
92 *Ibid.*, 68–70, J. Macoun to Editor of Statist, 24 March 1887
93 *Ibid.*, 85, J. Macoun to A. Waghorne, 28 September 1887
94 de Vecchi, 'The Pilgrim's Progress,' 521
95 NMCL, Macoun Letterbooks, vol. 1, 225, J. Macoun to T. Burgess, 16 January 1888
96 *Ibid.*, vol. 1, 259–60, 3 February 1888
97 The exact number of new species Macoun actually discovered is difficult to pin down. During his collecting heyday, he was often credited with finding several species new to science each field season; in 1887, for example, he evidently discovered fifty new species. Many of these species, however, were subsequently determined not to be new. A resolution of this question would entail a long, painstaking search of the original literature.
98 T. Pawlick, 'The Meadow Microcosm,' *Harrowsmith* (January 1981), 43–9, 140
99 NMNS, Botany Division, G. Vasey to J. Macoun, 6 September 1890

100 NMCL, Macoun Letterbooks, vol. 1, 177–8, J. Macoun to M.S. Bebb, 5 December 1887
101 NMNS, Botany Division, M.S. Bebb to J. Macoun, 6 February 1888
102 NMCL, Macoun Letterbooks, vol. 1, 212–13, J. Macoun to H.N. Ridley, 19 January 1888
103 *Ibid.*, 188, J. Macoun to G. Vasey, 10 December 1887
104 J. Macoun, *Catalogue of Canadian Plants, Part I – Polypetalae* (Montreal 1883); *Part II – Gamopetalae* (Montreal 1884); *Part III – Apetalae* (Montreal 1886)
105 *Botanical Gazette*, 12 (1887), 21
106 NMCL, Macoun Letterbooks, vol. 1, 85–6, J. Macoun to A. Waghorne, 28 September 1887
107 Dupree, *Asa Gray*, 390
108 Macoun, *Autobiography*, 254
109 Department of Interior, vol. 554, f. 165895, Thomas White to A.M. Burgess, 23 December 1887
110 *Ibid.*, draft of order-in-council approved by A.M. Burgess and signed by T. White

CHAPTER 3 The Search Widens

1 C. Berger, *Science, God and Nature in Victorian Canada* (Toronto 1983), 76–7; A.H. Dupree, *Asa Gray* (New York 1968), 393
2 NAC, Manuscript Division, MG 29, Robert Bell Papers, vol. 28, f. 4, 'Birds and Mammals'
3 J.B. Tyrrell, 'Catalogue of the Mammalia of Canada Exclusive of the Cetacea,' *Proceedings of the Canadian Institute*, 3rd series, 6 (October 1888), 66–91
4 J.F. Whiteaves, *Catalogue of Canadian Pinnipedia, Cetacea, Fishes and Marine Invertebrata* (Ottawa 1886)
5 NMCL, Macoun Letterbooks, vol. 2, 25, J. Macoun to E. Coues, 10 November 1888
6 NMNS, Botany Division, Ernest E.T. Seton to J. Macoun, 2 April 1885
7 *Ibid.*, J. Fletcher to James Melville [J.M.] Macoun, 20 July 1888
8 NMCL, Macoun Letterbooks, vol. 2, 50, J. Macoun to B.W. Evermann, 26 November 1888
9 Bell Papers, vol. 4, General Correspondence, 1886, J. Macoun to A.R.C. Selwyn, copy of letter in Bell's handwriting, nd (marginal note: 'Written before Colonial and Indian Exhibition')
10 NMNS, Botany Division, A.R.C. Selwyn to J.M. Macoun, 14 January 1887

11 NAC, Government Archives Division, RG 15, Department of Interior, vol. 554, f. 165895, P.B. Douglas to A.R.C. Selwyn, 15 March 1888

12 NAC, Government Archives Division, RG 45, GSC, Directors' Letterbooks, vol. 80, 112–17, A.R.C. Selwyn to J.A. Macdonald, 1885.

13 *Ibid.*, vol. 83, 424–6, 'Instructions for the Guidance and Preservation and Disposition of Various Birds, Mammals, etc. in the Museum,' March 1889

14 NAC, Manuscript Division, MG 30 B40, R.M. Anderson Papers, vol. 15, f. 9, M. Chamberlain to J. Macoun, 6 March 1885; M. Chamberlain to J.M. Macoun, 17 March 1886

15 *Ibid.*, M. Chamberlain to J. Macoun, 17 November 1885

16 *Ibid.*, 3 March 1886

17 *Ibid.*, 28 March 1888

18 GSC, Directors' Letterbooks, vol. 82, 444, A.R.C. Selwyn to M. Chamberlain, 14 December 1887

19 M. Chamberlain, *A Catalogue of Canadian Birds, with Notes on the Distribution of Species* (Saint John 1887), i

20 *Ibid.*, ii

21 NMCL, Macoun Letterbooks, vol. 1, 16, J. Macoun to A.R.C. Selwyn, 6 December 1885; vol. 1, 24, J. Macoun to W.E. Saunders, 21 January 1886

22 M.G. Ainley, 'From Natural History to Avian Biology: Canadian Ornithology, 1860–1950' (PHD diss., McGill University 1985), 74

23 NMCL, Macoun Letterbooks, vol. 1, 413, J. Macoun to J.P. Ellis, 18 February 1889

24 *Ibid.*, vol. 1, 358–9, J. Macoun to J. Vroom, 19 March 1888

25 *Ibid.*, vol. 1, 413, J. Macoun to M. Chamberlain, 4 April 1888

26 NAC, Anderson Papers, vol. 15, f. 9, M. Chamberlain to J. Macoun, 9 April 1888

27 NMNS, Botany Division, B.W. Evermann to J. Macoun, 3 December 1888

28 NMNS, Vertebrate Zoology Division, R.M. Anderson Papers, C.H. Merriam to J. Macoun, 25 May 1884

29 K.B. Sterling, *Last of the Naturalists: The Career of C. Hart Merriam* (New York 1976), 28

30 NMNS, Vertebrate Zoology Division, Anderson Papers, C.H. Merriam to J. Macoun, 7 June 1884, 5 June 1885, 21 July 1885

31 NMCL, Macoun Letterbooks, vol. 7, 139, J. Macoun to E. Britton, 20 February 1894

32 *Ibid.*, vol. 1, 450, J. Macoun to C.H. Merriam, 6 April 1888

33 D. Cole, *Captured Heritage: The Scramble for Northwest Coast Artifacts* (Seattle 1985), 74

34 For a discussion of Kennicott's and Baird's activities, see G. Thomas, 'The Smithsonian and the Hudson's Bay Company,' *Prairie Forum*, 10, no. 2 (Fall 1985), 283–305; W.A. Deiss, 'Spencer F. Baird and His Collectors,' *Journal of the Society for the Bibliography of Natural History*, 9, no. 4 (1980), 635–45; D. Lindsay, 'The Hudson's Bay Company-Smithsonian Connection and Fur Trade Intellectual Life: Bernard Rogan Ross, a Case Study,' in B.G. Trigger, T. Morantz, L. Dechêne, eds, *Le Castor fait tout* (Montreal 1987), 587–617

35 R.A. Jarrell, 'British Scientific Institutions and Canada: The Rhetoric and the Reality,' *Transactions of the Royal Society of Canada*, series iv, vol. 20 (1982), 537. For a list of the Canadian members of the AAAS for the period 1848–1900, see S.G. Kohlstedt, *The Formation of the American Scientific Community: The American Association for the Advancement of Science 1848–1860* (Chicago 1976)

36 Quoted in R.A. Proctor, 'Science in Canada,' *Knowledge*, 8 September 1882, 248

37 *Ibid.*, 247–8

38 Jarrell, 'British Scientific Institutions and Canada,' 533

39 P.J. Bowler, 'The Early Development of Scientific Societies in Canada,' in A. Oleson and J.C. Brown, eds, *The Pursuit of Knowledge in the Early American Republic* (Baltimore 1976), 332

40 NMCL, Macoun Letterbooks, vol. 2, 202–5, J. Macoun to T. McIlwraith, 28 February 1889

41 NMNS, Botany Division, T. McIlwraith to J. Macoun, 19 October 1887

42 NAC, Anderson Papers, T. McIlwraith to J. Macoun, 18 February 1889

43 NMCL, Macoun Letterbooks, vol. 2, 202–5, J. Macoun to T. McIlwraith, 28 February 1889

44 *Report of the Select Committee Appointed by the House of Commons to Obtain Information as to Geological Surveys* (Ottawa 1884), 185

45 P.A. Taverner, 'William Spreadborough – Collector, 1856–1931,' *Canadian Field-Naturalist*, 47, no. 3 (March 1933), 40

46 NMNS, Vertebrate Zoology Division, P.A. Taverner Papers, P.A. Taverner to E. Spreadborough, 19 January 1933

47 *Ibid.*

48 NMCL, Macoun Letterbooks, vol. 3, 32–5, J. Macoun to A.R.C. Selwyn, 2 January 1890

49 *Ibid.*, vol. 2, 304–5, J. Macoun to S. Watson, 16 September 1889

50 *Ibid.*, vol. 3, 9–10, J. Macoun to C.H. Merriam, 23 December 1889; vol. 3, 21, J. Macoun to C.H. Merriam, 30 December 1889; vol. 3, 83, J. Macoun to C.H. Merriam, 23 January 1890; vol. 4, 13–14, J. Macoun to C.H. Merriam, 28 March 1890

51 NMNS, Vertebrate Zoology Division, Anderson Papers, C.H. Merriam to J. Macoun, 2 April 1890
52 NMCL, Macoun Letterbooks, vol. 3, 182–4, J. Macoun to C.H. Merriam, 24 February 1890
53 A.R. Byrne, 'Man and Landscape Change in the Banff National Park Area Before 1911,' in R.C. Scace, ed., *Studies in Land Use History and Landscape Change*, National Park Series, no. 1 (April 1968), 131, 136–8; S. Van Kirk, 'Canada's Mountain National Parks and Federal Policy, 1885–1930' (MA diss., University of Alberta 1969); J. Foster, *Working for Wildlife: The Beginning of Preservation in Canada* (Toronto 1978), 29–30
54 NMCL, Macoun Letterbooks, vol. 5, 282–3, J. Macoun to E. Dewdney, 19 February 1892
55 NMNS, Vertebrate Zoology Division, Anderson Papers, William Spreadborough to J. Macoun, 29 May 1892. Unfortunately, Macoun would destroy Spreadborough's field notes after he had copied them into his bird catalogue manuscript
56 NMNS, Botany Division, J.M. Coulter to J. Macoun, 22 February 1888
57 J. Macoun, *Catalogue of Canadian Plants, Part IV – Endogens* (Montreal 1888); *Part V – Acrogens* (Montreal 1890). Endogens and acrogens are monocotyledons and ferns respectively
58 Quoted in *Ottawa Naturalist*, 5 (1891–2), 71–2
59 NMCL, Macoun Letterbooks, vol. 2, 157–8, J.M. Macoun to N.L. Britton, 2 February 1889
60 *Ibid.*, vol. 2, 1, J. Macoun to A.R.C. Selwyn, 18 December 1888
61 *Ibid.*, vol. 3, 127, J. Macoun to F.L. Walker, 11 February 1890
62 *Ibid.*, vol. 5, 471, J. Macoun to G.N. Best, 16 May 1892
63 *Ibid.*, vol. 3, 173–4, J. Macoun to E. Dewdney, 24 February 1890
64 *Ibid.*, vol. 5, 260–1, J. Macoun to G.M. Dawson, 9 February 1892
65 *Ibid.*, vol. 6, 31–2, J. Macoun to A.R.C. Selwyn, 14 March 1893
66 *Ibid.*, 369–70, J. Macoun to F.G. Wallbridge, 19 April 1893
67 NMNS, Botany Division, William Spreadborough to J. Macoun, 13 May 1892
68 *Ibid.*, E. Britton to J. Macoun, 15 February 1894. See also J.D. Godfrey, 'Notes on Hepaticae Collected by John Macoun in Southwestern British Columbia,' *Canadian Journal of Botany*, 50, no. 20 (1977), 2600–4; H. Crum, 'An Inventory of John Macoun's Musci,' *Occasional Papers of the Farlow Herbarium*, 16 (1981), 13–18
69 NMNS, Vertebrate Zoology Division, Anderson Papers, C.H. Merriam to J.M. Macoun, 17 January 1889
70 *Ibid.*, C.H. Merriam to J.M. Macoun, 4 May 1894
71 SIA, RU 161, Division of Reptiles and Amphibians, 1873–1968

Records, Box 29, folder 3, 454–7, L. Stejneger to J. Macoun, 29 March 1894

72 Canada, House of Commons, *Sessional Papers*, vol. 24, no. 17A, 1891, part iii, 'Geological Survey Department,' 54

73 NMNS, Botany Division, J.M. Macoun to T. Holm, 24 March 1910

74 GSC, Directors' Letterbooks, vol. 84, 329–30, Circular of Instructions, GSC, 25 April 1890

75 For a discussion of the Behring fur seal dispute, see R.C. Brown, *Canada's National Policy 1883–1900* (Princeton 1964)

76 NMCL, Macoun Letterbooks, vol. 4, 375, J.M. Macoun to J. Dearness, 5 February 1891; vol. 2, 188, J. Macoun to G. Kirkpatrick, 25 February 1889

77 *Ibid.*, vol. 5, 280–1, J. Macoun to M.S. Bebb, 9 February 1892

78 *Ibid.*, vol. 3, 260–7, J. Macoun to T. McIlwraith, 21 March 1889

79 NMCL, Macoun Letterbooks, vol. 4, 272–3, J. Macoun to N.L. Britton, 9 December 1890, 148–50; J. Macoun to N.L. Britton, 6 October 1890

80 *Ibid.*, 205–6, J. Macoun to T. McIlwraith, 5 January 1892

81 *Ibid.*, vol. 3, 101–4, J. Macoun to E. Britton, 31 January 1890

82 *Ibid.*, vol. 6, 102, J. Macoun to J. Fletcher, 3 December 1892

83 *Ibid.*, vol. 7, 218, J. Macoun to G.W. Taylor, 3 April 1894

84 *Ibid.*, vol. 4, 15, J. Macoun to S. Herring, 19 March 1890

85 *Ibid.*, vol. 5, 195, J. Macoun to W. Cross, 29 December 1891

86 *Ibid.*, 407–8, J. Macoun to W. Spreadborough, 10 May 1892

87 NMNS, Vertebrate Zoology Division, Taverner Papers, P.A. Taverner to E. Spreadborough, 12 January 1933; ROM, Percy A. Taverner Papers, P.A. Taverner to Fleming, 10 January 1933

88 NMCL, Macoun Letterbooks, vol. 7, 358, J.M. Macoun to T. Holm, 2 October 1894; vol. 8, 367, J.M. Macoun to T. Holm, 4 January 1895; S. Gwyn, *The Private Capital: Ambition and Love in the Age of Macdonald and Laurier* (Toronto 1984), 441

89 NMCL, Macoun Letterbooks, vol. 6, np, J.M. Macoun to J. Macoun, 17 July 1893

90 *Ibid.*

91 *Ibid.*, 388–9, J.M. Macoun to A.R.C. Selwyn, 20 July 1893

92 *Ibid.*, 358–60, J.M. Macoun to T. Holm, 2 October 1894

93 *Ibid.*, 63–5, J. Macoun to J.E. Bagnall, 24 September 1892

94 NMNS, Botany Division, J. Macoun to T. Holm, 29 December 1890

95 NMCL, Macoun Letterbooks, vol. 7, 187, J.M. Macoun to T. Holm, 18 March 1894

96 *Ibid.*, vol. 7, 340, J.M. Macoun to B.L. Robinson, 7 October 1894

97 *Ibid.*, vol. 1, 252–4, J.M. Macoun to M.S. Bebb, 22 February 1888

98 *Ibid.*, vol. 3, 171–2, J. Macoun to M.S. Bebb, 21 February 1890

99 *Ibid.*, vol. 1, 185–7, J. Macoun to N. Kindberg, 13 December 1887

100 *Ibid.*, vol. 4, 124–5, J. Macoun to W.H. Beeby, 30 September 1890

101 *Ibid.*, vol. 3, 162–3, J. Macoun to C.R. Barnes, 17 February 1890

102 *Ibid.*, vol. 4, 313, J. Macoun to G.N. Best, 8 January 1891

103 See Crum, 'An Inventory of John Macoun's Musci.'

104 NMCL, Macoun Letterbooks, vol. 4, 69–70, J. Macoun to E.G. Britton, 18 April 1890

105 *Ibid.*, 180–2, J. Macoun to E.G. Britton, 22 October 1890

106 NMNS, Botany Division, J.G. Bagnall to J. Macoun, 29 August 1892

107 *Ibid.*, E. Latnall to J. Macoun, 13 December 1892

108 *Botanical Gazette*, 17 (November 1892), 384–5

109 NMCL, Macoun Letterbooks, vol. 6, 185, J. Macoun to N. Kindberg, 22 December 1892

110 *Ibid.*, 486, J. Macoun to E.G. Britton, 30 October 1893

111 *Ibid.*, vol. 7, 84–5, J. Macoun to E.G. Britton, 15 January 1894

112 NMNS, Botany Division, E.G. Britton to J. Macoun, 11 January 1894; 15 January 1896; 21 March 1896

113 A.L. Andrews, 'John Macoun,' *Bryologist*, 24 (1921), 40

114 NMCL, Macoun Letterbooks, vol. 7, 182, J. Macoun to C. Bendire, 7 March 1894

115 NMNS, Vertebrate Zoology Division, Anderson Papers, C.H. Merriam to J. Macoun, 12 April 1892

116 NMNS, Botany Division, E.G. Britton to J. Macoun, 17 March 1896

117 NMCL, Macoun Letterbooks, vol. 7, 131, J. Macoun to S. Rhoads, 4 February 1894

118 *Ibid.*, 79–81, J. Macoun to C.H. Merriam, 4 January 1894

119 NMNS, Vertebrate Zoology Division, Anderson Papers, C.H. Merriam to J. Macoun, 28 January 1894

120 NMCL, Macoun Letterbooks, vol. 7, 118–20, J. Macoun to C.H. Merriam, 8 February 1894

121 NMNS, Vertebrate Zoology Division, Anderson Papers, S. Rhoads to J. Macoun, 4 December 1893

122 *Ibid.*, C.H. Merriam to J. Macoun, 4 December 1893

123 NMCL, Macoun Letterbooks, vol. 7, 211, J. Macoun to C.H. Merriam, 29 March 1894

124 *Ibid.*, 217, J. Macoun to S. Rhoads, 3 April 1894

125 NMNS, Vertebrate Zoology Division, Anderson Papers, C.H. Merriam to J. Macoun, 3 April 1894

126 NMCL, Macoun Letterbooks, vol. 7, 227–8, J. Macoun to C.H. Merriam, 5 April 1894

127 NMNS, Vertebrate Zoology Division, Anderson Papers, C.H. Merriam to J. Macoun, 2 May 1894
128 NMCL, Macoun Letterbooks, vol. 8, 103, J. Macoun to C.H. Merriam, 24 March 1896
129 GSC, Directors' Letterbooks, vol. 88, 547, A.R.C. Selwyn to T. Daly, 21 June 1894
130 NMCL, Macoun Letterbooks, vol. 7, 156, J.M. Macoun to C.G. Pringle, 28 February 1894
131 Ibid., vol. 7, 2–6, J. Macoun to A.R.C. Selwyn, 4 November 1893
132 Ibid., vol. 6, 264–5, J. Macoun to B. McClennan, 13 February 1893
133 Ibid., vol. 7, 118, J. Macoun to C.H. Merriam, 8 February 1894
134 Ibid., vol. 3, 148–9, J. Macoun to E.G. Britton, 14 February 1890
135 Ibid., vol. 4, 285, J. Macoun to N. Kindberg, 10 December 1890
136 Ibid., vol. 4, 331, J. Macoun to G.N. Best, 8 January 1891
137 Ibid., vol. 7, 374, J.M. Macoun to T. Holm, 16 October 1894

CHAPTER 4 Towards a National Museum

1 D.M. Knight, *The Age of Science: The Scientific World View in the Nineteenth Century* (New York 1986), 179
2 Canada, House of Commons, *Sessional Papers*, 1893, vol. 26, no. 13A, 'Summary Report of the Geological Survey of Canada for 1892,' 60
3 D.J. Hall, *Clifford Sifton, I: The Young Napoleon, 1861–1900* (Vancouver 1981), 125–7
4 *Ottawa Field-Naturalists' Club Transactions*, 1, no. 1 (1879–80), 6
5 *Ibid.*, 47
6 NMCL, Macoun Letterbooks, vol. 2, 276–7, J. Macoun to T.W. Burgess, 18 February 1888
7 *Ottawa Field-Naturalists' Club Transactions*, 6, no. 2 (1884–85), 176
8 House of Commons, *Debates*, 27 February 1877, 312
9 NAC, Manuscript Division, MG 26A, John A. Macdonald Papers, vol. 353, 163011–13, J.A. Grant to J.A. Macdonald, 21 December 1878
10 *Ibid.*, vol. 313, 141946–7, copy of Montreal Board of Trade report
11 House of Commons, *Debates*, 2 May 1882, 1266
12 *Ibid.*
13 While the transfer of the museum was being considered in 1878, James Grant had suggested to Prime Minister Macdonald that 'Stead's building on Sussex would be ideal.' Macdonald Papers, vol. 353, 163011–13, J.A. Grant to J.A. Macdonald, 21 December 1878
14 L. Brault, *The Mile of History* (Ottawa 1981), 65–7; *Bytown, A Guide to Lowertown* (Ottawa 1981), 40–1

15 Canada, House of Commons, *Sessional Papers*, 1882, vol. 15, no. 18, part iii, 'Geological and Natural History Survey,' 5–6

16 NAC, Government Archives Division, RG 45, GSC, Directors' Letterbooks, vol. 79, 262, A.R.C. Selwyn to D.L. Macpherson, 15 November 1883

17 Macdonald Papers, vol. 49, 112582–3, D.L. Macpherson to J.A. Macdonald, 19 April 1883

18 Canada, House of Commons, *Sessional Papers*, 1885, vol. 18, no. 13, part iii, 'Geological and Natural History Survey,' 29

19 *Ibid.*, 1887, vol. 20, no. 13, 'Annual Report of the Department of the Interior,' xxxvi

20 Ottawa *Citizen*, 10 April 1883, 4

21 J. Macoun, 'Presidential Address,' *Ottawa Naturalist*, 1, no. 11 (May 1887), 20

22 J.W. Dawson, 'Presidential Address,' *Transactions of the Royal Society of Canada*, series i, vol. 1 (1882–3), lvi

23 Macdonald Papers, vol. 83, 32329–3, Lord Lorne to J.A. Macdonald, 23 April 1883

24 *Transactions of the Royal Society of Canada*, series i, vol. 4 (1885–6), xxxi–xxxii

25 GSC, Directors' Letterbooks, vol. 84, 102, petition to A.R.C. Selwyn, 28 November 1889

26 Canada, House of Commons, *Sessional Papers*, 1883, vol. 16, no. 23, part ii, 'Geological and Natural History Survey,' 13

27 *Ibid.*, 1889, vol. 22, no. 15, part iii, 'Geological and Natural History Survey,' 40

28 H.M. Ami, 'Report on the State of the Principal Museums in Canada and Newfoundland,' *Report of the British Association for the Advancement of Science 1897* (London 1898), 62–74; S. Sheets-Pyenson, 'Better than a Travelling Circus: Museums and Meetings in Montreal during the Early 1880s,' *Transactions of the Royal Society of Canada*, series iv, vol. 20 (1982), 499–518; D. Cole, *Captured Heritage: The Scramble for Northwest Coast Artifacts* (Seattle 1985)

29 *Report of the Commissioners Appointed to Inquire into the State of the Public Records* (Ottawa 1898), 24

30 NAC, Manuscript Division, MG 30 B40, R.M. Anderson Papers, f. 9 ('Ornithology'), T. McIlwraith to J. Macoun, 1 March 1889

31 NMCL, Macoun Letterbooks, vol. 2, 260–3, J. Macoun to T. McIlwraith, 21 March 1889

32 *Ibid.*, vol. 2, 32–5, J. Macoun to A.R.C. Selwyn, 2 January 1890

33 GSC, Directors' Letterbooks, vol. 83, 449, Memorandum to Interior File 202831, 4 April 1889

34 'The National Museum,' *Canadian Mining Review* (February 1893), 11–12
35 'A National Museum Wanted,' *Journal of the General Mining Association of Quebec*, 1 (1893), 309
36 House of Commons, *Debates*, 28 June 1892, 439
37 'A National Museum Wanted,' 309
38 NAC, Government Archives Division, RG 11, Department of Public Works, vol. 4216, f. 688-1-A, T.M. Daly to J.A. Ouimet, 12 May 1893
39 *Ibid.*
40 M. Archibald, *By Federal Design: The Chief Architect's Branch of the Department of Public Works, 1881–1914* (Ottawa 1983), 36
41 NMCL, Macoun Letterbooks, vol. 7, 59, J. Macoun to M. Chamberlain, 22 December 1893
42 House of Commons, *Debates*, 30 April 1894, 2136
43 NMCL, Macoun Letterbooks, vol. 7, 59, J. Macoun to M. Chamberlain, 22 December 1893
44 MRBL, G.M. Dawson Papers, 1895 diary, introductory notes
45 M. Zaslow, *Reading the Rocks: The Story of the Geological Survey of Canada, 1842–1972* (Toronto 1975), 147–8
46 G.M. Dawson, 'The Progress and Trend of Scientific Investigation in Canada,' *Transactions Royal Society of Canada*, series i, vol. 12 (1894), lvi
47 The inventory listed 58,400 herbarium sheets, 2,300 bird skins, 693 mounted birds, 380 mammal skins, and 155 mounted mammals. NMCL, Macoun Letterbooks, vol. 8, 5–6, nd
48 *Ibid.*, vol. 8, 8, J. Macoun to W. Spreadborough, 12 February 1895
49 GSC, Directors' Letterbooks, vol. 89, 674–5, G.M. Dawson to T.M. Daly, 21 March 1895
50 *Ibid.*, vol. 90, 52–6, G.M. Dawson to T.M. Daly, nd
51 NMCL, Macoun Letterbooks, vol. 8, 17, J. Macoun to T. MacMillan, 16 February 1895
52 *Ibid.*, 83, J. Macoun to F.J. Washington, 1 April 1895
53 *Ibid.*, 143, J.M. Macoun to T. Holm, 20 May 1895
54 *Ibid.*, 338, J.M. Macoun to T. Holm, 14 December 1895
55 Canada, House of Commons, *Sessional Papers*, 1896, vol. 29, no. 13A, 'Annual Report of the Geological Survey for 1895,' 148
56 NAC, Manuscript Division, MG 24K, John Macoun Papers, 1895 field notebook, 19 May 1895
57 *Ibid.*, 20 May 1895
58 *Ibid.*, 29 July 1895
59 GSC, Directors' Letterbooks, vol. 90, 232, G.M. Dawson to T.M. Daly, 7 January 1896
60 *Ibid.*, vol. 92, 201–2, G.M. Dawson to H.J. Macdonald, 27 May 1896

61 NMCL, Macoun Letterbooks, vol. 9, 86, J.M. Macoun to E. Greene, 9 April 1896

62 C.S. Houston and M.J. Bechard, 'Early Manitoba Oologists,' *Blue Jay*, 45, no. 2 (June 1987), 85–96

63 Canada, House of Commons, *Sessional Papers*, 1897, vol. 30, no. 13A, 'Annual Report of the Geological Survey for 1890,' 136

64 NMCL, Macoun Letterbooks, vol. 9, 115, J. Macoun to E. Britton, 27 April 1896

65 Archibald, *By Federal Design*, 36

66 GSC, Directors' Letterbooks, vol. 92, 553, G.M. Dawson to R.W. Scott, 9 September 1896

67 Canada, House of Commons, *Sessional Papers*, 1897, vol. 30, no. 13A, 'Summary Report of the Geological Survey for 1896,' 6A

68 GSC, Directors' Letterbooks, vol. 93, 362, G.M. Dawson to C. Sifton, 6 January 1897

69 House of Commons, *Debates*, 17 May 1897, 2430–2

70 *Ibid.*, 2432

71 *Ibid.*, 2433

72 Department of Public Works, vol. 4216, f. 688-1-A, D. Scott to J.I. Tarte, 26 May 1897

73 Hall, *Clifford Sifton, I*, 125–7

74 Canada, House of Commons, *Sessional Papers*, 1897, vol. 30, no 13A, 'Annual Report of the Geological Survey for 1896,' 3; Zaslow, *Reading the Rocks*, 202

75 NMCL, Macoun Letterbooks, vol. 9, 376–7, J. Macoun to W. Spreadborough, 6 April 1897

76 *Ibid.*, 18–19, J. Macoun to N. Kindberg, 4 November 1896

77 Canada, House of Commons, *Sessional Papers*, 1898, vol. 32, no. 13A, 'Annual Report of the Geological Survey for 1897,' 146

78 'Annual Report of the Geological Survey for 1896,' 125

79 *Ibid.*, 10–11

80 GSC, Directors' Letterbooks, vol. 97, 181, G.M. Dawson to N.A. Belcourt, 20 March 1898

81 *Report of the Commissioners Appointed to Inquire into the State of the Public Records*, 24

82 House of Commons, *Debates*, 8 June 1898, 7478–81

83 MRBL, Dawson Papers, 1898 diary, 18 October 1898

84 GSC, Directors' Letterbooks, vol. 98, 502–3, G.M. Dawson to C. Sifton, 31 October 1898

85 *Canadian Record of Science*, 7, no. 3 (1896), 199

86 'A National Museum,' *Canadian Mining Review* (April 1897), 4

87 Quoted in W. Eggleston, *The Queen's Choice: A Story of Canada's Capital* (Ottawa 1961), 154–5

88 *Transactions of the Royal Society of Canada*, series ii, vol. 5 (1899), xvii

89 House of Commons, *Debates*, 26 April 1899, 2106–7

90 GSC, Directors' Letterbooks, vol. 100, 240, G.M. Dawson to N.A. Belcourt, 1 June 1899

91 Canada, House of Commons, *Sessional Papers*, 1899, vol. 33, no. 13A, 'Annual Report of the Department of the Interior for 1898,' 3

92 *Ibid.*, 195

93 GSC, Directors' Letterbooks, vol. 100, 337, G.M. Dawson to J. Macoun, 17 June 1899

94 MRBL, Dawson Papers, 1897 diary, 14 December 1897

95 NAC, Government Archives Division, RG 2, Privy Council Office, PC 1723, 30 June 1898

96 GSC, Directors' Letterbooks, vol. 102, 552–3, G.M. Dawson to W. Mulock, 6 April 1900

97 For the 1899–1900 fiscal year, the Department of Public Works spent $568,874.90 in the Yukon district. Canada, House of Commons, *Sessional Papers*, 1901, vol. 35, no. 1, part v, 'Auditor General's Report,' 2

98 GSC, Directors' Letterbooks, vol. 100, 510, G.M. Dawson to N.A. Belcourt, 19 July 1899

99 UTL, B.E. Walker Papers, B.E. Walker to G.M. Dawson, 29 July 1899

100 GSC, Directors' Letterbooks, vol. 101, 250, G.M. Dawson to D. Ewart, 2 October 1899

101 *Ottawa Naturalist*, 13 (1899–1900), 274–5

102 B.E. Walker, 'Canadian Surveys and Museums and the Need of Increased Expenditure Thereon,' *Proceedings of the Canadian Institute*, 3rd series (1899), 88

103 NAC, Manuscript Division, MG 26G, Wilfrid Laurier Papers, vol. 156, 45880–2, G.M. Dawson to W. Laurier, 23 May 1900

104 J. Macoun, *Catalogue of Canadian Birds, Part I. Water Birds, Gallinaceous Birds and Pigeons* (Ottawa 1900)

105 *Auk*, 17, no. 17 (October 1900), 395

106 NMNS, Vertebrate Zoology Division, Anderson Papers, C.H. Merriam to J. Macoun, 30 June 1900

107 NMNS, Botany Division, J. Fletcher to J. Macoun, 24 September 1900

108 NMCL, Macoun Letterbooks, vol. 6, 167, J. Macoun to W. Spreadborough, 7 December 1892

109 *Ibid.*, vol. 9, 580–3, J. Macoun to G. Ross, 28 June 1900

110 Department of Public Works, vol. 4216, f. 688-1-A, J.I. Tarte to

A. Gobeil, 8 February 1901 (trans: 'Please place in the supplementary estimates a sum of $50,000 for the Geology Museum.')
111 GSC, Directors' Letterbooks, vol. 105, 61, G.M. Dawson to C. Sifton, 14 February 1901
112 *Ibid.*, 122, G.M. Dawson to J.I. Tarte, 27 February 1901
113 Canada, House of Commons, *Sessional Papers*, 1901, vol. 35, no. 26, 'Annual Report of the Geological Survey for 1900,' 190

<h3 style="text-align:center">CHAPTER 5 Canada's Century</h3>

1 M. Zaslow, *Reading the Rocks: The Story of the Geological Survey of Canada, 1842-1972* (Toronto 1975), 217
2 *Ibid.*, 216
3 *Ibid.*, 208–10
4 *Ibid.*; NAC, Government Archives Division, RG 72, Canadian Government Exhibition Commission, vol. 37, f. 155, R. Bell to C. Tupper, 29 November 1885; NAC Manuscript Division, MG 26A John A. Macdonald Papers, vol. 428, pp. 209355–6, R. Bell to J.A. Macdonald, 24 July 1886
5 NAC, Manuscript Division, Clifford Sifton Papers, vol. 49, R. Bell to C. Sifton, 15 April 1898
6 House of Commons, *Debates*, 19 May 1888, 1605–6
7 NAC, Government Archives Division, RG 45, GSC, Directors' Letterbooks, vol. 84, 124–6, A.R.C. Selwyn to E. Dewdney, 18 December 1890
8 An Act Respecting the Department of the Geological Survey, 53 Victoria, ch. 2, 1890
9 GSC, Directors' Letterbooks, vol. 92, 95, G.M. Dawson to J.W. Tyrell, 29 April 1896
10 D.J. Hall, *Clifford Sifton, II: The Lonely Eminence, 1901–1929* (Vancouver 1985), 49–50
11 Sifton Papers, vol. 49, C. Sifton to R. Bell, 6 March 1901; Zaslow, *Reading the Rocks*, 241–5
12 House of Commons, *Debates*, 25 April 1901, 3770
13 Zaslow, *Reading the Rocks*, 215
14 Canada, House of Commons, *Sessional Papers*, 1902, vol. 36, no. 19, 'Summary Report of the Geological Survey for 1901,' 150
15 Canada, Senate, *Journals*, 1888, vol. 22, no. 1, 'Report of the Select Committee of the Senate appointed to Inquire into the Resources of the Great Mackenzie Basin,' 247
16 *Ibid.*, 243–4
17 'Summary Report of the Geological Survey for 1901,' 190

18 P.A. Taverner, 'William Spreadborough – Collector 1856–1931,' *Canadian Field-Naturalist*, 47, no. 3 (March 1933), 41

19 Zaslow, *Reading the Rocks*, 209–10

20 'Summary Report of the Geological Survey for 1901,' 154

21 *Ottawa Naturalist*, 15, no. 1 (February 1902), 239–44; NMNS, Vertebrate Zoology Division, Anderson Papers, C.H. Merriam to J.M. Macoun, 11 February 1902

22 Canada, House of Commons, *Sessional Papers*, 1903, vol. 37, no. 26, 'Summary Report of the Geological Survey for 1902,' 50

23 Canada, House of Commons, *Journals*, 1903, vol. 38, App. 2, 'Report of the Select Standing Committee on Agriculture and Colonization,' 61

24 Zaslow, *Reading the Rocks*, 216; Sifton Papers, vol. 49, C. Sifton to R. Bell, 17 July 1902

25 House of Commons, *Debates*, 20 May 1901, 5682–3

26 Major's Hill Park had first been suggested as a possible site for the national museum building by former Liberal Minister of the Interior, David Laird. *Ibid.*, 28 June 1892, 4390

27 *Ibid.*, 28 February 1902

28 NAC, Manuscript Division, MG 26G, Wilfrid Laurier Papers, vol. 247, 68841–8, petition to Laurier, nd

29 *Auk*, 22 (January 1905), 99

30 NMNS, Botany Division, P.A. Taverner to J. Macoun, 21 December 1904

31 *Ibid.*, J.M. Macoun to J.E. Carrier, 2 October 1904

32 *Ibid.*, J.R. Anderson to J.M. Macoun, 1 February 1906

33 House of Commons, *Debates*, 23 March 1903, 366; 17 October 1903, 14244

34 C.J. Taylor, *Some Early Ottawa Buildings* (Ottawa 1975), 369, 374–5

35 House of Commons, *Debates*, 24 October 1904, 401

36 Zaslow, *Reading the Rocks*, 212

37 GSC, Directors' Letterbooks, vol. 108, 488, R. Bell to C. Sifton, 31 March 1904

38 House of Commons, *Debates*, 17 July 1903, 6817

39 Sifton Papers, vol. 255, C. Sifton to R. Bickerdike, 26 January 1904

40 Canada, House of Commons, *Journals*, 1904, vol. 39, App. 2, 'Report of the Select Standing Committee on Agriculture and Colonization,' 433

41 NMNS, Botany Division, J. Macoun to J.C. Schultz, November 1888

42 Today, there are 4.5 million acres under cultivation in the Peace, with approximately 900,000 acres each of wheat, barley, and canola. For the early-twentieth-century agricultural history of the region, see C.A. Dawson and R.W. Murchie, *The Settlement of the Peace River Country: A Study of a Pioneer Area* (Toronto 1934).

43 J.M. Macoun, *Report on the Peace River Region* (Ottawa 1904), 5–6, 40
44 Sifton Papers, vol. 257, C. Sifton to R. Bell, 2 April 1901
45 GSC, Directors' Letterbooks, vol. 108, 489, R. Bell to C. Sifton, 5 April 1904
46 Sifton Papers, vol. 257, C. Sifton to H.T. Irwin, 7 April 1904
47 *Ibid.*, C. Sifton to R. Bell, 8 April 1904
48 *Ibid.*
49 GSC, Directors' Letterbooks, vol. 108, 504, R. Bell to C. Sifton, 11 April 1904
50 H.Y. Hind, *Manitoba and the North-West Frauds* (Windsor 1883), 23
51 'Report of the Select Standing Committee on Agriculture and Colonization,' 1904, 340
52 *Ibid.*, 355
53 *Ibid.*
54 House of Commons, *Debates*, 11 April 1904, 988
55 D.J. Hall, *Clifford Sifton, I: The Young Napoleon, 1861–1900* (Vancouver 1981), 282–3
56 Quoted in *Ibid.*, 296
57 Sifton Papers, vol. 168, F. Oliver to C. Sifton, 30 March 1904. For a discussion of this issue, see M. Zaslow, 'A History of Transportation and Development of the Mackenzie Basin from 1871 to 1921' (MA diss., University of Toronto 1948), ch. 3
58 House of Commons, *Debates*, 15 April 1904, 1311–18
59 Zaslow, *Reading the Rocks*, 255
60 *Daily Edmonton Bulletin*, 6 April 1904, 2
61 *Ibid.*
62 *Ibid.*, 8 April 1904, 2
63 *Ibid.*, 12 April 1904, 2
64 House of Commons, *Debates*, 5 April 1904, 785
65 'Report of the Select Standing Committee on Agriculture and Colonization,' 1904, 435–9
66 *Daily Edmonton Bulletin*, 6 April 1904, 2; 22 April 1904, 2; 30 April 1904, 2
67 NMNS, Botany Division, W. Spreadborough to J.M. Macoun, 9 May 1904
68 'Report of the Select Standing Committee on Agriculture and Colonization,' 1904, 433
69 *Ibid.*, 478–9. Macoun's verbal battle with Oliver on 14 April had been reported in several newspapers. See, for example, *Manitoba Free Press*, 15 April 1904, 12; Toronto *Globe*, 15 April 1904, 4; Ottawa *Morning Citizen*, 15 April 1904, 4; Montreal *Gazette*, 15 April 1904, 5
70 Sifton Papers, vol. 258, C. Sifton to R. Bell, 29 April 1904; NMNS, Botany Division, J.M. Macoun to W. McInnis, 28 March 1916

71 'Report of the Select Standing Committee on Agriculture and Colonization,' 1904, 512
72 *Ibid.*, 549
73 NMNS, Botany Division, J.M. Macoun to C.B. Burns, 25 April 1911
74 'Report of the Select Standing Committee on Agriculture and Colonization,' 1904, 434
75 House of Commons, *Debates*, 20 July 1904, 7201
76 *Ibid.*, 7204
77 *Ibid.*, 7207
78 *Ibid.*, 7350
79 'Report of the Select Standing Committee on Agriculture and Colonization,' 1904, xiv
80 M. Zaslow, *The Opening of the Canadian North, 1870–1914* (Toronto 1971), 209
81 House of Commons, *Debates*, 20 July 1904, 7209
82 NMNS, Botany Division, J.M. Macoun to W. Spreadborough, 3 June 1904
83 House of Commons, *Debates*, 20 July 1904, 7201, 7211
84 *Ibid.*, 19 July 1904, 7062
85 NMNS, Botany Division, J. Macoun to A.H. Mackay, 20 May 1903
86 *Ibid.*, E.L. Greene to J.M. Macoun, 31 December 1904
87 *Ibid.*, Botany Division, J.M. Macoun to T. Holm, 7 April 1910; Canada, House of Commons, *Sessional Papers*, 1905, vol. 39, no. 26, 'Summary Report of the Geological Survey for 1904,' 269
88 NMNS, Botany Division, J.N. Rose to J.M. Macoun, 12 July 1906
89 NMNS, Vertebrate Zoology Division, Anderson Papers, C.H. Merriam to J. Macoun, 16 February 1906
90 *Ibid.*, Taverner Papers, J. Macoun to C.H. Merriam, 2 March 1906
91 *Ibid.*, Anderson Papers, C.H. Merriam to J. Macoun, 13 March 1906
92 House of Commons, *Debates*, 9 February 1905, 750
93 Canada, House of Commons, *Sessional Papers*, 1906, vol. 40, no. 26, 'Summary Report of the Geological Survey for 1905,' 144; 'Summary Report of the Geological Survey for 1904,' 384
94 House of Commons, *Debates*, 12 July 1905, 9304–5
95 Laurier Papers, vol. 375, 99901–3, R. Bell to W. Laurier, 22 July 1905
96 *Ibid.*, 99904, W. Laurier to R. Bell, 27 July 1905
97 *Ibid.*, vol. 407, 108079–81, F. Oliver to W. Laurier, 13 March 1906
98 GSC, Directors' Letterbooks, vol. 112, 157, R. Bell to F. Adams, 2 February 1906
99 House of Commons, *Debates*, 22 June 1906, 5942

CHAPTER 6 The Natural History Division

1 M. Zaslow, *Reading the Rocks: The Story of the Geological Survey of Canada, 1842–1972* (Toronto 1975), 264–6
2 W. Coleman, *Biology in the Nineteenth Century: Problems of Form, Function and Transformation* (Cambridge 1977), 2–3
3 NAC, Government Archives Division, RG 45, GSC, Directors' Letterbooks, vol. 112, 233, A.P. Low to C.D. Walcott, 4 April 1906
4 *Ibid.*, 208–9, A.P. Low to F. Oliver, 31 March 1906
5 Canada, House of Commons, *Sessional Papers*, 1906–7, vol. 41, no. 6, 'Summary Report of the Geological Survey for 1906,' 172
6 NMNS, Botany Division, 1906 field notebook, 12 July 1906
7 J. Macoun, 'On the Explorations along the Grand Trunk Pacific Railway Between Portage la Prairie and Edmonton,' in *Sessional Papers*, 1906–7, vol. 41, no. 6, 81
8 Canada, House of Commons, *Journals*, 1906–7, vol. 42, App. 4, 'Report of the Select Standing Committee on Agriculture and Colonization,' 1
9 *Ibid.*, 8–9
10 *Ibid.*, 24 (emphasis added)
11 NMCL, Macoun Letterbooks, vol. 2, 72, J.M. Macoun to B. Everman, 6 December 1888; vol. 4, 457, J.M. Macoun to C.E. Bendire, 3 April 1891
12 NAC, Manuscript Division, MG 26G, Laurier Papers, vol. 403, 107370, J. Macoun to F. Oliver, 1 February 1906
13 *Ibid.*, vol. 40, 106673–4, J. Macoun to W. Laurier, 9 February 1906
14 House of Commons, *Debates*, 22 June 1906, 5983
15 Zaslow, *Reading the Rocks*, 257–8
16 An Act to Create a Department of Mines, 6–7 Edward VII, ch. 29, 1907
17 GSC, Directors' Letterbooks, vol. 114, 337, A.P. Low to J. Macoun, 8 July 1907
18 *Ibid.*, 523–4, P. Selwyn to T.M. Daly, 15 November 1907
19 *Ibid.*, 519–20, P. Selwyn to R. Roblin, 8 November 1907
20 A. Gibson, 'C.H. Young 1867–1940,' *Canadian Field-Naturalist*, 55 (February 1941), 20–1
21 Young's summary produced 2,302 bird skins and 1,106 mammal skins
22 NMNS, Botany Division, J.M. Macoun to T. Holm, 4 February 1910
23 NAC, Government Archives Division, RG 11, Department of Public Works, vol. 4217, f. 688-1-B, George Goodwin to C.S. Hyman, 22 August 1905
24 Montreal *Gazette*, 13 October 1973
25 Zaslow, *Reading the Rocks*, 287
26 *Ibid.*, 264–6

27 *Ibid.*, 279
28 NMNS, Botany Division, J.M. Macoun to W. Spreadborough, 10 January 1908
29 *Ibid.*
30 Canada, House of Commons, *Sessional Papers*, 1909, vol. 43, no. 26, 'Summary Report of the Geological Branch for 1908,' 189
31 Coleman, *Biology in the Nineteenth Century*, 2–3; D.M. Knight, *The Age of Science: The Scientific World View in the Nineteenth Century* (New York 1986), 4–7
32 NMNS, Botany Division, J. Macoun to E. Britton, 6 December 1909
33 *Ibid.*, J. Macoun to W.H. Dall, 7 November 1910
34 *Ibid.*, J.M. Macoun to P.A. Taverner, 2 February 1910
35 J. Macoun and J.M. Macoun, *Catalogue of Canadian Birds* (Ottawa 1909), iv–v
36 ROM, J.H. Fleming Papers, P.A. Taverner to J.H. Fleming, 20 January 1910
37 NMNS, Botany Division, J.M. Macoun to W.E. Saunders, 15 December 1909
38 *Ibid.*, W. Spreadborough to J. Macoun, 17 January 1910
39 *Ibid.*, J.M. Macoun to N.L. Britton, 19 March 1918
40 NMNS, Vertebrate Zoology Division, Anderson Papers, J.M. Macoun to C.H. Merriam, 28 April 1910
41 NMNS, Botany Division, J.M. Macoun to H. Bartlett, 6 February 1911
42 NMCL, Macoun Letterbooks, vol. 6, 224, J. Macoun to [recipient not named], 20 January 1893
43 NMNS, Botany Division, J.M. Macoun to R.W. Brock, 15 February 1911
44 *Ibid.*, J. Macoun to E. Britton, 17 January 1911
45 House of Commons, *Debates*, 23 March 1911, 5937–8
46 Laurier Papers, vol. 610, 105795–9, C. Sifton to W. Laurier, 3 December 1909
47 *Ibid.*, 165793–4, W. Templeman to W. Laurier, 25 January 1910
48 *Ibid.*, 165979–80, petition to W. Laurier, 2 December 1910
49 House of Commons, *Debates*, 23 March 1911, 5937–8
50 Zaslow, *Reading the Rocks*, 279
51 NMNS, Botany Division, J. Macoun to G.U. Hay, 9 February 1911
52 Canada, House of Commons, *Sessional Papers*, 1911, vol. 45, no. 26, 'Summary Report of the Geological Survey Branch for 1910,' 2, 9
53 ROM, Fleming Papers, J.H. Fleming to P.A. Taverner, 6 October 1906, 16 October 1906
54 NMNS, Botany Division, W.E. Saunders to J. Macoun, 14 April 1910
55 *Ibid.*, J.M. Macoun to P.A. Taverner, 18 April 1911

56 ROM, Fleming Papers, P.A. Taverner to J. H. Fleming, 13 May 1911
57 NMNS, Botany Division, J.M. Macoun to C.B. Burns, 25 April 1911
58 *Ibid.*, J.M. Macoun to E.L. Greene, 4 April 1911
59 *Ibid.*, J.M. Macoun to M.L. Fernald, 1 June 1906
60 *Ibid.*, J.M. Macoun to E.L. Greene, 6 April 1911
61 *Ibid.*, J.M. Macoun to T. Holm, 6 April 1911
62 The manuscript was turned over to Dr Brock but never printed. It is currently housed in a storage room at the Botany Division, National Museum of Natural Sciences, Ottawa
63 ROM, Fleming Papers, P.A. Taverner to J.H. Fleming, 10 June 1911
64 NMNS, Botany Division, J. Macoun to E.C. Allen, 17 January 1911
65 *Ibid.*, J. Macoun to J. Dearness, 3 July 1911
66 Canada, House of Commons, *Sessional Papers*, 1912, vol. 46, no. 22, 'Summary Report of the Geological Survey Branch for 1911,' 373
67 A.C. Gluek, 'Canada's Splendid Bargain: The North Pacific Fur Seal Convention of 1911,' *Canadian Historical Review*, 63, no. 2 (June 1982), 179–201
68 NMNS, Botany Division, J.M. Macoun to E.L. Greene, 23 October 1911
69 *Toronto Star Weekly*, 10 December 1925, 35
70 NMNS, Vertebrate Zoology Division, P.A. Taverner to R.W. Brock, 12 May 1911
71 M. Ainley, 'From Natural History to Avian Biology: Canadian Ornithology, 1860–1950' (PHD diss., McGill University 1985), 115–16
72 NMNS, Vertebrate Zoology Division, P.A. Taverner to R.W. Brock, 12 May 1911
73 ROM, Fleming Papers, P.A. Taverner to J.H. Fleming, 13 May 1911
74 'Summary Report of the Geological Branch for 1911,' 399
75 NMNS, Vertebrate Zoology Division, Taverner Papers, P.A. Taverner to R.W. Brock, 27 July 1911
76 'Summary Report of the Geological Survey Branch for 1911,' 374
77 ROM, Fleming Papers, P.A. Taverner to J.H. Fleming, 7 September 1911
78 NMNS, Botany Division, J. Macoun to C.M. Fraser, 13 February 1912

CHAPTER 7 The End of an Era

1 Mary Macoun Kennedy interview, 27 January 1976
2 Eleanor Sanderson interview, 30 April 1980
3 NMNS, Vertebrate Zoology Division, Taverner Papers, P.A. Taverner to W.T. Macoun, 23 March 1921
4 *Ibid.*, P.A. Taverner to J. Macoun, 20 May 1912
5 The purchase of Taverner's 1,200 specimen collection was justified on

the grounds that it consisted of eastern birds, whereas the Survey collection was chiefly made up of western specimens. NMNS, Botany Division, J. Macoun to R.W. Brock, 12 February 1912. For a list of those collections that were purchased for the new museum by Taverner, see the Victoria Memorial Museum Register of Birds

6 P.A. Taverner, *On the Collection of Zoological Specimens for the Victoria Memorial Museum* (Ottawa 1912)

7 NMNS, Vertebrate Zoology Division, Taverner Papers, P.A. Taverner to B.H. Swales, 23 October 1912

8 *Ibid.*, P.A. Taverner to J. Macoun, 20 May 1912

9 *Ibid.*, J.M. Macoun to P.A. Taverner, 2 July 1912

10 Canada, House of Commons, *Sessional Papers*, 1913, vol. 47, no. 26, 'Summary Report of the Geological Survey Branch for 1912,' 441

11 H. Lloyd, 'Clyde Louis Patch,' *Canadian Field-Naturalist*, 68 (July–September 1954), 124–6

12 M. Zaslow, *Reading the Rocks: The Story of the Geological Survey of Canada, 1842–1972* (Toronto 1975), 319–22

13 J.D. Soper, 'In Memoriam: Rudolph Martin Anderson, 1876–1961,' *Canadian Field-Naturalist*, 76 (July–September 1962), 127–33

14 NMNS, Botany Division, J.M. Macoun to R.W. Brock, 12 March 1913

15 *Ibid.*, R.W. Brock to J. Macoun, 16 April 1913

16 *Ibid.*, J.M. Macoun to R.W. Brock, 26 May 1913

17 NAC, Government Archives Division, RG 2, Privy Council Office, PC 1387, 9 June 1913

18 NMNS, Botany Division, J.M. Macoun to R.W. Brock, 10 November 1913

19 NMNS, Vertebrate Zoology Division, Taverner Papers, J. Macoun to P.A. Taverner, 7 March 1915

20 NMNS, Botany Division, J. Macoun to R.W. Brock, 9 January 1914

21 Canada, House of Commons, *Sessional Papers*, 1914, vol. 48, no. 26, 'Summary Report of the Geological Survey Branch for 1913,' 344

22 NMNS, Botany Division, J. Macoun to R.W. Brock, 9 January 1914

23 Canada, House of Commons, *Sessional Papers*, 1915, vol. 49, no. 26, 'Summary Report of the Geological Survey Branch for 1914,' 151

24 NMNS, Botany Division, museum committee memorandum to R.W. Brock, 16 October 1914

25 *Ibid.*

26 Zaslow, *Reading the Rocks*, 307, 310

27 NMNS, Vertebrate Zoology Division, J. Macoun to P.A. Taverner, 7 March 1915

28 Canada, House of Commons, *Sessional Papers*, 1916, vol. 51, no. 26, 'Summary Report of the Geological Survey Branch for 1915,' 251

29 NMNS, Botany Division, museum committee memorandum to R.W. Brock, 16 October 1914

30 Zaslow, *Reading the Rocks*, 310

31 NMNS, Botany Division, J.M. Macoun to R.G. McConnell, 19 October 1916

32 *Ibid.*, R.G. McConnell to J.M. Macoun, 28 February 1918

33 *Ibid.*, J.M. Macoun to M.L. Fernald, 20 February 1918

34 *Ibid.*, J.M. Macoun to R.G. McConnell, 3 July 1919

35 *Ibid.*, R.M. Anderson to J.M. Macoun, 15 July 1919

36 *Ibid.*, J.M. Macoun to J. Marshall, 7 March 1916

37 NMNS, Vertebrate Zoology Division, Anderson Papers, R.M. Anderson to W. Spreadborough, 13 November 1919, 17 November 1919

38 *Ibid.*, Taverner Papers, P.A. Taverner to A. Brooks, 22 January 1920

39 *Ibid.*, Anderson Papers, R.M. Anderson to W. Spreadborough, 13 November 1919

40 ROM, Percy A. Taverner Papers, P.A. Taverner to J.H. Fleming, 10 January 1933

41 NMNS, Vertebrate Zoology Division, Anderson Papers, J. Macoun to M.O. Malte, 13 March 1920

42 *Ibid.*, J. Macoun to R.M. Anderson, 27 March 1920

43 *Ibid.*

44 Originally entitled 'The Evolution of a Naturalist,' the autobiography was dictated largely from memory to a local secretary. This arrangement did not make it any easier for Macoun, who would often tire of answering questions or trying to remember details; he often stormed out of 'Ninety-Eight' to the refuge of his little cottage across the street. NMNS, Vertebrate Zoology Division, Anderson Papers, R.M. Anderson to W.T. Macoun, 26 January 1920

45 *Ibid.*, J. Macoun to R.M. Anderson, 5 July 1920

46 Ottawa *Citizen*, 31 July 1920

47 Zaslow, *Reading the Rocks*, 353–6

CONCLUSION

1 NMNS, Botany Division, M.O. Malte to C. Camsell, 20 December 1924

2 Canada, House of Commons, *Journals*, 1906–7, vol. 42, App. 4, 'Report of the Select Standing Committee on Agriculture and Colonization,' 18

3 NMNS, Botany Division, J.M. Macoun to T. Holm, 7 April 1910
4 *Ibid.*, M.O. Malte to C. Camsell, 20 December 1924
5 For a list of species named after Macoun, see J. Macoun, *Autobiography of John Macoun, M.A.*, second edition (Ottawa 1979), 304–5, 327–8
6 *Transactions of the Royal Society of Canada*, series iii, vol. 16 (1922), xxi

Index